THE EVOLUTION OF SILICON VALLEY

A HISTORY IN PICTURES

BY MARY WADDEN

In 2011, venture capitalist John Doerr & his wife Ann hosted a dinner for President Obama at their Woodside home. The intimate guest list was a who's who of Silicon Valley featuring the head of technology companies including Apple, Cisco, Facebook, Genentech, Google, Netflix, Oracle, Twitter, and Yahoo! as well as the President of Stanford University. The White House via Wikimedia.

MIX
Paper | Supporting responsible forestry
FSC® C208677
FSC www.fsc.org

This book is an expanded sequel to ***Silicon Valley: The History in Pictures*** *which was originally published in 2013.*

ISBN: 978-0-578-44636-3

First Edition, Third Printing

Printed in China using FSC® - Certified and other controlled material.

Questions? Comments? mary@siliconvalleysmarts.com

Foreword

Born and raised in the Bay Area, I became curious as to why this particular place—once known merely for its temperate climate and fertile soil—has spawned the microprocessor, popularized the personal computer, given rise to the Internet and created social media. Why did these inventions originate here and not anywhere else? While researching the answer to this question, I uncovered a rich and complex history. Looking through old photos, the story came alive. Once I embarked on this project, I became acutely aware of why I could write this book now but could not have conceived of it even 30 years ago. My three pound Macbook Air possesses the processing power of all of the government computers running in the 1960s...combined! The Adobe Suite of design applications provided me with publishing tools once only reserved for multinational corporations. When I found myself confronted with questions like, 'What is the difference between an integrated circuit and a microprocessor?' Google helped me navigate to the answer. Silicon Valley continues to transform our day to day lives. This book is my tribute to the region where the only constant is change itself.

Acknowledgments

Many people supported me in the research and design of this book and its predecessor Silicon Valley: The History in Pictures.

I would specifically like to thank: my mother and sister, Rosemary and Elizabeth Wadden, for providing design and visual help as well as all those who encouraged me to take this project on, especially Dick Genest. This book would not have been possible at all if it wasn't for the numerous organizations and corporations who provided images, insight and wisdom to complete this book, including but not limited to:

* The Computer History Museum & The Santa Clara Valley Historical Association
* Stanford Special Collections and University Archives
* The Historical Societies of Redwood City, Menlo Park, Palo Alto, Los Altos, Mountain View, Sunnyvale, Santa Clara, Saratoga, Milpitas and San Mateo County
* SJSU Special Collections & The California Room, San Jose Public Library, Moffett Field Museum, San Lorenzo Valley Museum, Sempervirens Club, The Iron Man Museum, the San Jose Airport Authority & the NASDAQ archives
* Companies including AMD, Applied Materials, Cisco, Cypress Semiconductor, Fetch Robotics, FMC, Ford Land Co, IBM, Intel, Intuit, Juniper Networks, LinkedIn, Lockheed Martin, Nvidia, Space Systems Loral, SRI, Synopsis, Testarossa Winery, Texas Instruments, TiVo, Trimble, Xerox PARC, & View, Inc.
* Individual photo contributors including Pat Charles, Gaye Clemson, Janice Cuban, Donna Dubinsky, Martin Eberhard, Trip Hawkins, Gary Hendrix, Bob Lash, Max Levchin, Jamis MacNivin, T.J. Rodgers, Sebastian Thrun & Robyn Twomey.

Contents

WARNING!
NOTICE IS GIVEN that any person found Pilfering, Stealing, Robbing, or committing any act of Lawless Violence will be summarily
HANGED
BEST CHOICE
BRAND
CALIFORNIA PEARS

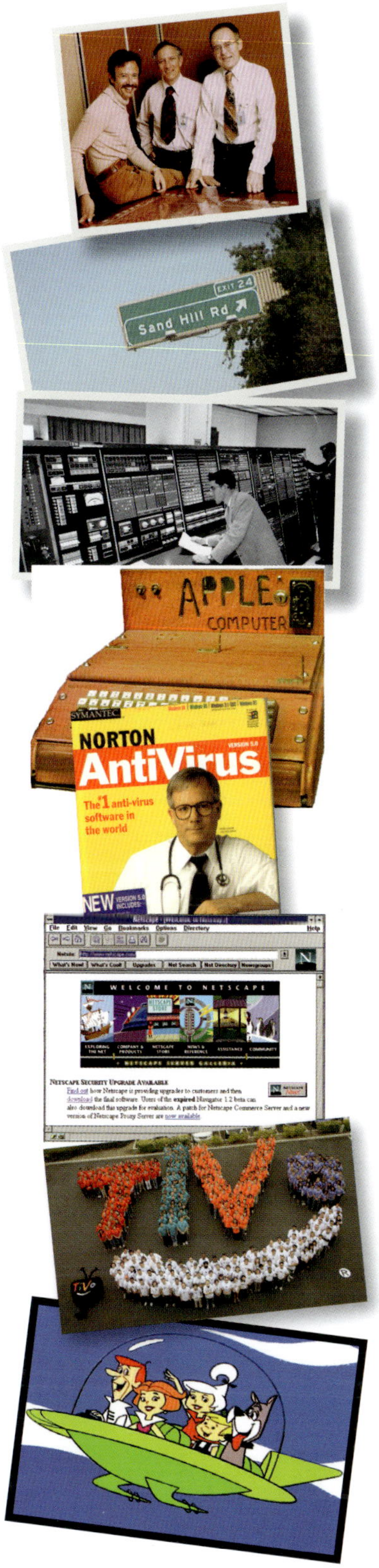
EXIT 24
Sand Hill Rd
APPLE
COMPUTER
NORTON
AntiVirus
The #1 anti-virus software in the world
NEW
TiVo

Downtown Saratoga, late 1800s. *Saratoga Historical Society.*

The modern age of California began in the mid 1700s with the arrival of Spanish missionaries. The following 100 years were a transformative time for Santa Clara Valley as the land essentially changed hands three separate times during that period: first from the native Ohlone population to the Spanish Missionaries, then from the Missionaries to Mexican Rancheros and finally from the Rancheros to the 49'ers who arrived in search of gold. This era was marked by a tremendous influx of settlers and was truly the Wild West.

One: The Wild West

Mission Santa Clara

Mission Santa Clara today, in the middle of Santa Clara University. Wikimedia

In the sixteenth century, Spain conquered the Aztec civilization in present day Mexico and spent the next 300 years in a quest for empire in the New World. In 1769 a group of Spanish Missionaries led by Junipero Serra set out to continue the line of missions into Alta California. Between 1769 and 1823, twenty one missions were established as a way to gain empire for the crown of Spain. Legend has it that the route between the missions was named El Camino Real meaning "The Royal Way," as a tribute to the Spanish royalty who financed the quest. However, this piece of folklore is now generally considered more of a myth. The Mission of Santa Clara was established in 1777 as the eighth mission and was originally constructed near the Guadalupe River. It has been destroyed and rebuilt five times. Prior to the arrival of the missionaries, the Ohlone people lived near the bay for thousands of years. Unlike the Aztecs and Mayans, the Ohlone didn't build large cities but lived in small settlements. They survived by hunting and gathering, living a peaceful life in harmony with their environment. With the establishment of the Santa Clara Mission, the friars imposed a new life on the natives: a new religion, a new language, new survival skills (including farming and raising cattle) along with a host of new diseases. More natives were buried at the Santa Clara Mission than at any of the other missions.

Pueblo San Jose

Map of Downtown San Jose from 1886. Downtown San Jose developed around the initial pueblo which was near what is now the intersection of Highways 280 and 87. Library of Congress

A Spanish frontier colony typically had three parts: the mission, the presidio and the pueblo. Missions were established to convert the native population; presidios were military forts; and pueblos were civil settlements tasked with providing food for the presidios. El Pueblo de San José was founded in 1777 a few months after Mission Santa Clara and the two institutions were located about three miles apart and connected by The Alameda. The Pueblo de San Jose was responsible for providing food and other supplies for the San Francisco and Monterey Presidios. The first census of the Pueblo of San Jose was in 1831 and totaled 524 inhabitants.

Mexican land Grants & The Resulting Ranchos

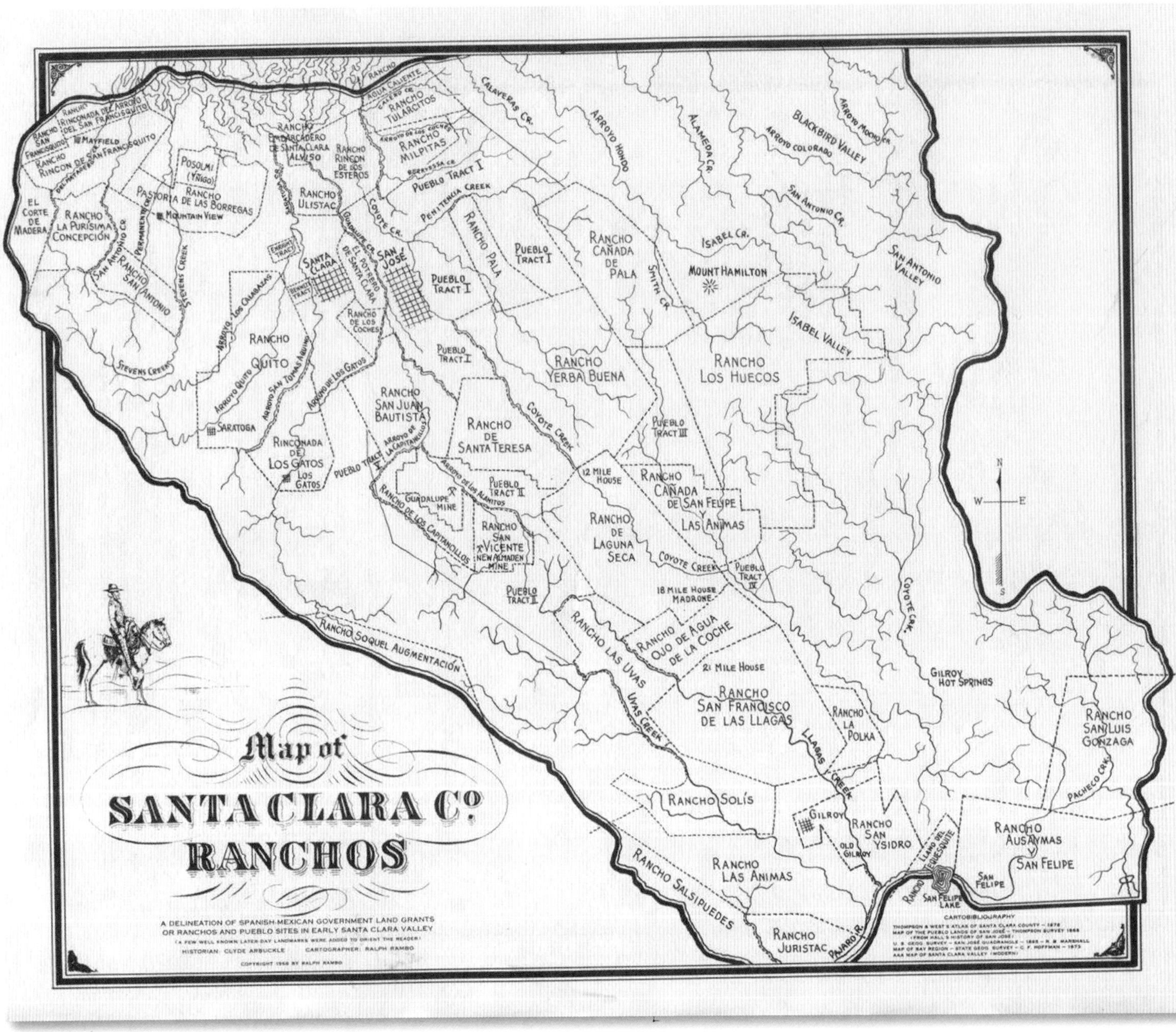

When Mexico declared their independence from Spain, the Mexican government divided the lands once held by the missions and awarded Ranchos to private citizens. Library of Congress.

The Jose Maria Alviso adobe house, built in 1837 and located in present day Milpitas, is now a historical monument as an example of the architecture from the days of the Ranchos.

In 1821 Mexico declared its independence from Spain and a California Assembly was created. The Assembly issued directives to put the land that was under control of the Spanish missions into secular hands. Under Mexican rule, the area was known as Alta California and the people who lived here were "Californios." The Santa Clara Mission claimed control of land from current day Palo Alto through Gilroy. In 1834 more than 800 ranchos were awarded in Santa Clara Valley ranging from a house lot up to 50,000 acres. Thus, most California rancheros became land barons to a greater or lessor degree. Authors have often referred to this period of history as the "Days of the Dons" and according to the Palo Alto Historical Association: *These rancho days were famed for beautiful senoritas, daring and handsome lovers, brave hunters, brutal sports and lavish hospitality*. Many cities, neighborhoods and streets got their names from this era of history. Adobe architecture was the most prevalent during the rancho days. The photograph above is of an adobe house built in 1837 by Jose Maria Alviso who was an early alcalde (mayor) of the Pueblo de San Jose. The home was built in the Monterey style which features a wood-shingled roof, wood balconies on three sides, French doors, interior fireplaces, and a symmetrical layout. The Alviso Adobe contains a remarkable amount of its original features; adobe walls from the 1830s, windows, doors and hardware from 1853 and an almost intact 1920s kitchen.

The Gold Rush

Posters advertising transport to California during the early 1850s. *The Bancroft Library*

The great California gold rush began in 1848 when a gold nugget was discovered in the American River. News of the discovery brought thousands of immigrants to California from all over the world. At first, there were only two ways to travel out West. The first entailed a six month sea voyage from New York around the tip of South America to California. This journey brought with it rampant seasickness, bug infested food and a high price tag. The second route brought travelers in covered wagons over the rugged terrain and hostile territory of the Oregon-California Trail which also took six months. By 1850 the difficulty of both routes spurred construction of the Panama Railway, which shaved months off the voyage and helped foster continued immigration out west. The large influx of '49ers, as the gold prospectors were known, caused California's population to increase exponentially. Between 1848–1850, the population of San Francisco grew from 1,000 to over 20,000.

Uncomfortable conditions on the stage coach line prompted Wells Fargo to post the following rules for passenger behavior:

- *Abstinence from liquor is requested, but if you must drink share the bottle. To do otherwise makes you appear selfish and unneighborly.*
- *If ladies are present, gentlemen are urged to forgo smoking cigars and pipes as the odor of same is repugnant to the gentler sex.*
- *Chewing tobacco is permitted, but spit with the wind, not against it.*
- *Don't snore loudly while sleeping or use your fellow passenger's shoulder for a pillow.*
- *Firearms may be kept for use in emergencies. Do not fire them for pleasure or shoot at wild animals as the sound riles the horses.*
- *Forbidden topics of conversation include stagecoach robberies and Indian uprisings.*

San Jose Serves as the First State Capital

In 1899 San Jose built a replica of the first state capitol building as shown here, to mark the 50th anniversary of the event.

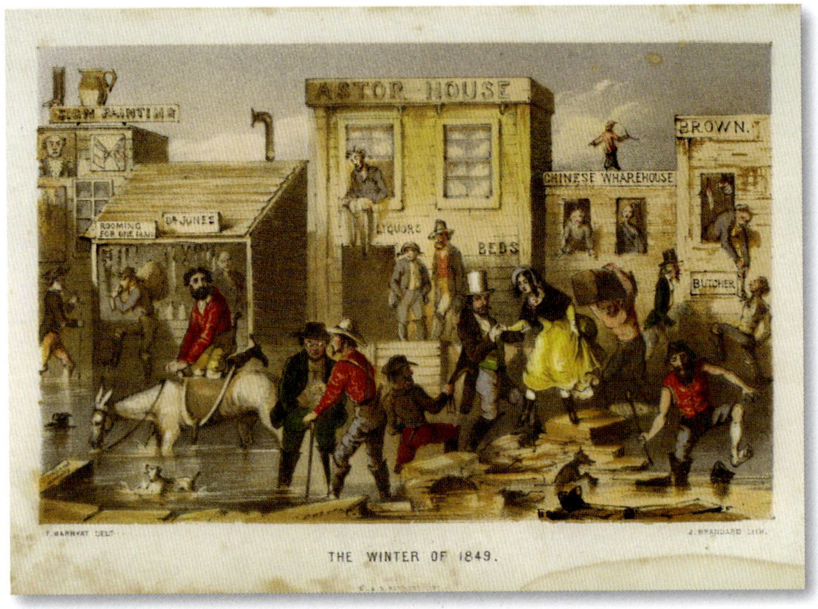

A postcard depicting the wet and rather haphazard conditions the legislators found themselves in during the winter of 1849. The Bancroft Library.

In 1849 California's elected officials met in Monterey to draft a constitution in hopes that California would be admitted into the United States. After the people of California approved the new constitution it was sent to the United States Government. It took nearly a year for Congress to deliberate over the question of admitting California into the Union. Finally, and perhaps not coincidentally after gold had been discovered in California, President Fillmore signed the documents declaring California the 31st state. The first and second legislative sessions were held in San Jose in a two story adobe hotel, no more than 40 x 60 feet, during one of the wettest winters on record.

"Legislature of 1000 Drinks"

Palace Hotel Saloon, located at the corner of Cahill Street and The Alameda. This photo was taken in 1902 and is one of the only images of a bar in San Jose during this era. According to San Jose historian, Leonard McKay, the second story of the saloon was a brothel.

It rained 36 inches the winter of 1849, well over twice the annual average for San Jose. The lawmakers complained about the accommodations and about the inclement weather and resorted to spending most of their time in the local saloons. Senator Green of Sacramento is credited for encouraging his fellow lawmakers with the slogan "Let's have a drink! Let's have a thousand drinks!" Due to the poor experience in San Jose, the capital was moved to Vallejo the following year and then permanently to Sacramento in 1854.

Cupertino Winery circa 1891. California History Center.

Many who came West in the hopes of striking it rich mining for gold soon found a better life was available to homesteaders who produced the necessities for California's burgeoning population. With its abundance of natural resources, fertile soil and temperate weather, Santa Clara Valley and the Peninsula became the ideal spot for farmers, lumbermen ,and viticulturists. The abundant yield of the wheat crops grown in the fertile valley, the massive amounts of lumber produced in the Santa Cruz Mountains and the wine grapes that grew perfectly in the temperate climate of the foothills soon earned the area the title of "The Eden of the World."

Two: Early Industries

California's Majestic Redwoods

Reaching heights over 320 feet (the equivalent of a 30 floor skyscraper), California's state tree is in fact the world's tallest. The Coast Redwood can grow a trunk as wide as 24 feet and can live more than 2,000 years. The native population of California did not cut these trees down, but used fallen wood to build shelter and canoes. However, as hundreds of thousands of people flooded into California after the Gold Rush in 1850, the need for housing and timber grew exponentially and many looked to the Giant Redwood as the solution. Lumber from the Santa Cruz mountains became a hugely lucrative business for early Californians and the resulting deforestation went unregulated for the next 50 years.

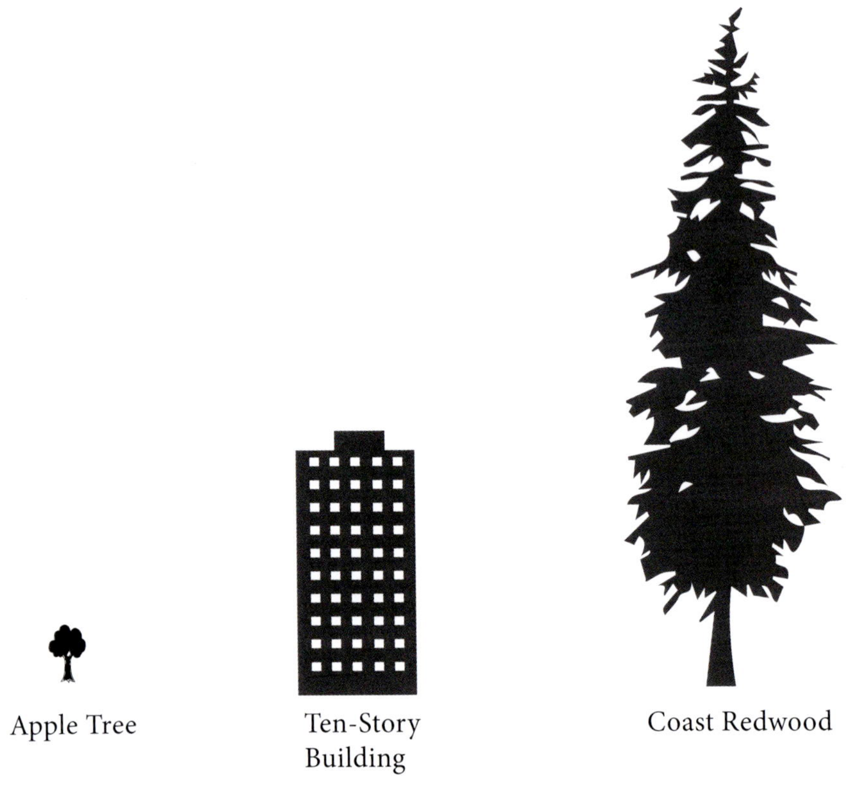

Tourists posing within one of the giant redwoods in the Santa Cruz mountains circa 1880. San Lorenzo Valley Museum

1905.
Geo. L.
CAMP 4

LUMBERMEN

The lumbermen who arrived in the Santa Cruz mountains brought with them a long single blade called a whipsaw to cut the wood. The lumbermen would either dig a large pit or construct a sturdy platform where one man, the "pit-man," was positioned below the log and the "top-man" stood on top of the platform. The two-man team worked together to raise, lower, and guide the saw. The pit-man had to contend with sawdust in his mouth and eyes, not to mention the risk of being crushed by a falling log. While it was sometimes risky, the process tedious and the life lonely, many men found they could make more money splitting lumber than by working in the gold mines.

Whipsawers circa 1880.

Early lumbermen posing in the mountains circa 1880. San Lorenzo Valley Museum

Lumber Transportation

Oxen carrying the lumber down the mountainside, circa 1880. San Lorenzo Valley Museum.

Initially, lumbermen depended on longhorn oxen to carry the weight of the logs to the mills. The first step was to build a skid road which followed natural gullies downhill to the mill. They would bury logs every few feet in the skid road to help secure it and prevent unwanted mudslides. The final step was to smear beef tallow on each buried log, making it easier for the oxen to drag the cut logs down. Eight to ten yoke of oxen were used to pull a train of logs down the skid road to the mill.

South Pacific Coast Railway crossing over the Los Gatos Creek Trail, circa 1892. San Lorenzo Valley Museum.

Tunnel near Scotts Valley, circa 1892. San Lorenzo Valley Museum.

The introduction of the South Pacific Coast Railroad, which served the Santa Cruz Mountains beginning in the 1880s, made for faster and easier transportation of lumber. The newfound efficiency of the railroad coupled with the increased power of new mechanical logging methods caused lumber production to increase tenfold resulting in vast deforestation. In 1880 alone, Santa Cruz County generated ten million board feet of lumber.

Lumber Mills

One of the largest lumber mills at the time was the Pacific Manufacturing Company which was established in Santa Clara in 1874 and shortly thereafter became the largest wood product supplier on the Pacific Coast. Their lumber yard alone covered twenty acres and the mill warranted a private switch for the Southern Pacific Railroad. At the height of production the company employed up to 600 people. The business closed in 1960.

The Pacific Manufacturing plant circa 1898. *Santa Clara City Library Heritage Pavilion*

ANY. LUMBER YARD.
PACIFIC MANUFACTURIN
LUMBER, MILL WORK, DOORS, SASH, BLIND

Restoration of The Forests

In 1899 San Jose photographer, Andrew P. Hill, was on assignment in the Santa Cruz Mountains. He had been sent to gather images of a fire in the redwoods that had been put out using wine from a local vineyard. The experience had a great effect on Hill and shortly thereafter, he began a quest to save the redwoods from complete deforestation. Hill convened a meeting with Stanford University's president, David Starr Jordan, as well as prominent scientists and other influential people to discuss what could be done to protect the redwoods. Out of this meeting came the Sempervirens Club whose mission was to lobby for the protection of the redwoods. After three years of campaigning, the efforts of the Sempervirens Club paid off. The state acquired 3,800 acres and created Big Basin Redwoods State Park, the first park established in California. Today, only 5% of the original redwoods remain and are all located in State Parks. This means that 95% of the redwood trees that currently exist today sprouted in the last 150 years after the deforestation of the late nineteenth century. This is especially sad when considering a redwood tree's potential 2000 year life span.

The first meeting of the Sempervirens Club in 1899. Sempervirens Club ▶

Preservation acts by groups like the Sempervirens Club have helped restore much of the natural beauty to the California redwood forests as seen here.

SEMPERVIRENS
CLUB.

GRAIN

Harvest time in 1910. Silicon Valley Historical Association

In the early 1850s farmers in the Santa Clara Valley realized that even without fertilizer, the land produced the best yields on wheat and barley they had ever seen. Thus began the period of commercial farming in the Valley. Wheat continued to be the dominant crop grown in the Valley from 1850–1880 and by 1868 the Santa Clara Valley was growing enough wheat to feed the entire state. During the Valley's wheat days, ranches sold for around $20 per acre. After orchards took over the landscape in the twentieth century, land was worth upwards of $400 per acre.

Farmers bailing hay circa 1890. Silicon Valley Historical Association

Postcard of Forbes Flour Mill in Los Gatos, circa 1905. California Room

Hay Stacks along the Alameda in San Jose. California Room

Wine

The history of wine-making and viticulture in the Santa Clara Valley goes back to the 18th Century when the padres at the Santa Clara Mission grew grapes for ceremonial wine. Testarossa Winery

The Los Gatos Saratoga Wine and Fruit Company was built by local growers in 1885 to facilitate the processing of grapes. The winery was located on Saratoga-Los Gatos Road (Highway 9) in present day Monte Sereno. Saratoga Historical Foundation

In 1852 Antoine Delmas imported his first European grapevines into the state. Eight years later, he also imported the first European winery and distillery equipment. Between 1849–1852 Santa Clara County attracted thousands of French and Italian settlers who immediately realized the potential of the valley and foothills for creating a world class wine district. At the time, wine was sold by the barrel to community grocery stores where customers simply brought their own bottles and filled them directly from the wine barrels. By 1870 the region's wines were consistently winning awards.

Wine

The Sacred Heart Novitiate in Los Gatos. Testarossa Winery

In 1871 the president of Santa Clara University acquired acreage in the West Valley to produce wine for the school. Known as Villa Maria, this successful enterprise paved the way for the Jesuit Sacred Heart Novitiate and Winery in Los Gatos in the 1880s. Vincent Pichetti arrived from Italy in 1872 and managed the Villa Maria wine-making operation for more than a decade, later acquiring his own vineyard which is still around today. Each autumn from 1888 to 1986, the Jesuit winery harvested grapes. At one time, they picked enough fruit to produce 150,000 gallons of wine per year.

The Sacred Heart Novitiate in Los Gatos after prohibition. Boxing up wine grapes, potentially to send off to the East Coast. Circa 1920. Testarossa Winery

With the ratification of the Eighteenth Amendment to the United States Constitution in 1919, the era of Prohibition began and with it wiped out the burgeoning wine industry in Santa Clara Valley. Almost all of the wineries in California closed their doors between 1920 and 1933. However, during the period of Prohibition people were still allowed to make wine at home for personal use—to the tune of 200 gallons of wine per year. Due to this rather interesting provision, demand for California's wine grapes remained high. For this reason wine grape production remained consistent even as commercial wine-making all but disappeared. Wine making in Valley began again in the 1940s. Today, the Santa Cruz Mountains Winegrowers' Association boasts 48 member wineries. Most are small, family-owned operations and a number of them still have ties to the early wine-makers of the area.

Beer

The Santa Clara Brewery, photographed here in 1895, was owned by the German American Lauck family. It was located at the northeast corner of Benton and Alviso Streets. Santa Clara City Library

Local beer making was another industry that arose from the arrival of the '49ers. There were a few notable characters who originally came west to make their fortunes in the gold mines but later returned to their known trade from back home. In 1853 the first brewery in the Santa Clara Valley was established and was named the Eagle Brewery. It was started by Joe Hartman in a shack on South Market Street in San Jose and in its humble beginnings, if a saloon needed a keg of beer, Joe put the keg in his wheelbarrow and delivered it himself. As business flourished, he moved to the corner of San Carlos and Market Streets and built the Eagle Brewery where the St. Claire Hotel stands today. The malt tower of the Eagle Brewery had six stories; making it the tallest building in San Jose at the time. In 1869 competition arrived when a German tavern keeper started the Fredericksburg Brewery on the corner of The Alameda and Cinnabar Street which is featured on the following spread. By the 1870s, a number of new breweries had popped up to accommodate the growing population. While the barley used was grown locally, the hops, bottles and barrels had to arrive by rail so most breweries were located on the railway.

Fredericksburg Brewery circa 1884.

Fredericksburg Brewery circa 1895.

The Fredericksburg Brewery opened in 1869 at the corner of The Alameda and Cinnabar Street by a German immigrant. It was one of the most popular businesses in the area until it closed in 1918. The following is a first-hand account of a brewery tour found in *Pen Pictures from the Garden of the World*, published by the Pacific Press Publishing Company in 1888:

> *"Embarking on one of the handsome cars of the Electric Road we are whirled rapidly along the famous Alameda Avenue, with its leafy shade, past the homes of wealthy men, sheltered with giant trees and embowered in flowers, to a point where stirring life and bustling activity proclaim the presence of some great enterprise. It is the Fredericksburg Brewery, the widest known and the most extensive establishment of its kind west of the Rocky Mountains."*

Franklin Street in downtown Santa Clara in 1898. *Santa Clara City Library Heritage Pavilion*

As logging and agriculture continued to flourish, towns began to spring up to support the growing population. The condition of affairs in San Jose around 1850 was described by Grandma Bascom who migrated to the area from Kansas. The following is an excerpt from *Overland Monthly and Out West Magazine*, May 1887:

> *Once in San Francisco we hoped to find a house which the Doctor had bought in New York and ordered sent around the Horn. He had also sent in the same cargo a great lot of furniture and a year's supply of provisions, but they didn't come until the next April and then everything was spoiled but the house…We always intended to come to the Santa Clara Valley, for the Doctor said that wherever the Catholic Fathers had picked out a site it must be a good one. The children and I stayed in the city while the Doctor went on horseback to San Jose and bought a house for us… We came to Alviso in the boat and paid $150 in fare, just for me and the children. From Alviso we came to San Jose by the Pioneer stage through fearful mud and pouring rain, paying an ounce* (of gold) *for each fare.* (Once in San Jose), *people began to ask if they could stay with us until they found some other home…The Legislature was in session and the town was more than full. Before I knew it, I had thirteen boarders—senators and representatives, ministers and teachers…In 1852 we moved out to our own farm in Santa Clara. I remember we paid our head carpenter $16 a day. The house cost us $10,000. It would not cost $1,000 now. An ounce of onion seed cost an ounce of gold. We paid $6 each for our fruit trees. A mule cost $300; a horse $400. But at the time, doctor's services were just as high-priced and so we kept even.*

Three: Towns

San Jose

Postcard of Downtown San Jose from 1920.

Downtown San Jose in 1908. History San Jose.

Postcard of Downtown San Jose in 1896. This view is looking north down market street with St. Joseph Church on the right.

In 1775 Juan Bautista de Anza led the first overland expedition to bring colonists from New Spain (Mexico) to Alta California to prepare for two missions, one presidio (military fort), and one pueblo (town). The site of the pueblo was initially located near the present-day intersection of Guadalupe Parkway and Taylor Street and was designed as a farming community to serve the presidios of San Francisco and Monterey. In 1797 the pueblo was moved from its original location to what is now Plaza de César Chávez in downtown San Jose. In 1881 San Jose installed a 237 foot electrified tower as an alternative to lighting the city with gas. The experiment didn't go as planned but the tower remained operational for years as an attraction to be lit up on holidays and festivals. See postcard lower left.

Santa Clara

Franklin Street in downtown Santa Clara in 1898. *Santa Clara City Library Heritage Pavilion.*

In the 1850s Santa Clara began to take shape as a recognizable small town. The town site was surveyed by William Campbell into lots of one hundred square yards. One lot was given to each citizen with the understanding that he was to build a house on it within three months or lose the property. A schoolhouse and a church were built; several hotels erected; mercantile businesses established; and 23 houses were imported from Boston to be set up in the town. In 1851 Santa Clara University was established on the old mission site and became a prominent feature of the developing town. Two of the earliest manufacturing businesses in Santa Clara were the Wampach Tannery, established in 1849, and the Pacific Manufacturing Company, established in 1874.

Los Gatos

Main Street in Los Gatos looking east circa 1908. Looking closely, you can see a gato crossing in front of the stagecoach. Saratoga Historical Society.

Los Gatos received its name due to the many mountain lions that came to drink from the Los Gatos Creek. In 1854 James Alexander Forbes purchased some land on the Los Gatos Creek and built a flour mill, which became the nucleus around which the town of Los Gatos was built. Los Gatos became a desired place for city dwellers to build a summer home, as well as a resting spot for railroad passengers traveling between San Francisco and Santa Cruz.

Postcard of Los Gatos, circa 1910.

LEXINGTON / ALMA

Alma Train Station, circa 1925.

The Alma Country Store.

1915 Alma view across the meadow towards the country store.

Train station in Alma, circa 1925.

In 1860 an immigrant from Kentucky bought some property at the base of the Santa Cruz Mountains and named it Lexington after his home town. This little outpost became a stop on the stagecoach route from Los Gatos to Santa Cruz. The local base of influence transferred from Lexington to another outpost nearby named Alma, when in 1880, the narrow gauge railway from Alviso to Santa Cruz went through Alma instead of Lexington. Folklore says the town of Alma was named after a local prostitute. Alma had a hotel, saloons and a general merchandise store. Following the completion of State Route 17, the railroad ceased operations in 1940 and the area fell into a period of decline. In 1952 the Lexington Reservoir was created and both Lexington and Alma were officially abandoned. Old roads and building foundations from these frontier towns are often visible under Lexington Reservoir during periods of low water level.

Saratoga

Saratoga post office circa 1894. Saratoga Historical Society.

Saratoga derived its name in the 1850s from the discovery of mineral springs, which had a chemical content similar to the great Saratoga Springs in New York. The first acknowledged permanent American settler in Saratoga was William Campbell who started a sawmill operation in 1847. Saratoga soon became a stagecoach stop between the Santa Cruz Mountains and the valley floor. Hotels and saloons were strung out along Lumber Street (now Big Basin Way) and a rough frontier atmosphere prevailed as late as the 1880s.

Postcard of the Saratoga Foothills.

Cupertino

Cupertino Store in 1939 at the corner of Stevens Creek and Highway 9. California History Center.

The village of Cupertino sprang up at the crossroads of Highway 9 (now De Anza Boulevard) and Stevens Creek Road in the foothills of the valley. It was first known as West Side but by 1898 the post office at the crossroads needed a new name to distinguish it from other similarly named towns. John T. Doyle, a San Francisco lawyer and historian, had given the name Cupertino to his winery. In 1904 the name Cupertino caught on and was made official.

Postcard of the rolling foothills of Santa Clara County.

SUNNYVALE

Murphy Avenue in Downtown Sunnyvale, circa 1915. Sunnyvale Historical Society.

Sunnyvale was originally named after Martin Murphy Jr, the first prominent resident in the area. Murphy Jr. was part of the Stephens-Townsend-Murphy Party, famous for being the first party to cross the Sierras which they accomplished in 1844. In 1850 Martin Murphy Jr. purchased 4,800 acres for about $1 an acre and soon became the most wealthy man in California. In 1860 Murphy allowed The San Francisco and San Jose Railroad to establish Murphy Station and later Lawrence Station on his property. In 1901 the residents of Murphy were informed that they could not use the name Murphy for their post office and decided to use the name Sunnyvale, not just for the post office but for the name of their town as well. The Murphy residence as seen to the right was a clapboard house built in Maine, disassembled, shipped around Cape Horn to Sunnyvale and then reassembled. In 1881 Martin Murphy Jr. celebrated his golden wedding anniversary and invited the entire state of California. Attendance to the party was estimated to be between 10,000 - 15,000 people.

The Murphy House. Sunnyvale Historical Society.

ALVISO

The Alviso Hotel in 1905. California Room.

In 1840 Ignacio Alviso moved south from Mission Santa Clara to the Rancho Rincon de Los Esteros, meaning corner of the estuaries. The town that bears his name was once a thriving port, which offered merchants a way to transport their beaver pelts, cattle hides and tallow by ship to San Francisco. Alviso was instrumental in establishing both the port and a fine hotel pictured here. Some buildings from that era still exist today.

The Alviso Regatta during its heyday. South Bay Yacht Club.

Mountain View

Castro Street looking south. City of Mountain View.

In 1842 Francisco Estrada and his wife, Inez Castro were granted the 8,800 acre Rancho Pastoria de las Borregas. In 1850 a Stage Coach stop was established near Grant Road and El Camino Real. Naturally, the initial commercial hub of Mountain View developed in this area. In the 1860s, Henry Rengstorff built docks, wharves and warehouses near present day Shoreline Amphitheatre to ship the region's agricultural products to San Francisco. The San Francisco and San Jose Railroad was completed in 1864 and placed a depot on Evelyn Avenue, which shifted the downtown area from Grant Road to Castro Street.

Postcard of the oak trees that were scattered along the valley floor.

Los Altos

Downtown Los Altos, circa 1900. Los Altos History Museum.

In the 1800s, most of the acreage that is now Los Altos and Los Altos Hills was included in two Mexican land grants, Rancho la Purissima Concepcion and Rancho San Antonio. In the late 1800s, Sarah Winchester owned the 100 acres that later became the town site of Los Altos. In 1906 the Southern Pacific Railroad bought the land from Winchester for a right of way for their trains. Paul Shoup, who later became president of the Southern Pacific Railroad, helped create the town when he and others formed the Altos Land Company. The Shoup family built several summer homes along University Avenue that are still there today.

Map of the land purchased by the railroads from Sarah Winchester.

MAYFIELD

Mayfield's Main Street looking northwest from Sheridan Street circa 1886. Palo Alto Historical Association.

Uncle Jim's Cabin, a.k.a. the tavern of James Otterson, taken in 1874. Palo Alto Historical Association.

The township of Mayfield was formed in 1855 and grew up around James Otterson's hotel, which opened in 1853 at the corner of El Camino Real and Lincoln Avenue (which is now known as California Avenue). Known as Uncle Jim's Cabin, it was patronized by travelers en route between San Francisco and San Jose and by lumbermen driving down from the hills. In the later part of the 19th century, Mayfield embodied the Wild West town spirit with over a dozen saloons and perhaps half as many brothels. However, by 1925 Mayfield decided to become part of Palo Alto due to economic considerations, which explains why Palo Alto has two downtown areas: one along University Avenue and one along California Avenue.

Palo Alto

Looking down University Avenue in 1911 with an approaching street car. Silicon Valley Historical Association.

The city got its name from the tall landmark redwood tree, El Palo Alto, which still grows on the east bank of San Francisquito Creek across from Menlo Park. In 1886 Leland Stanford met with leaders of Mayfield to discuss his plans for the university and the need for a nearby town. He had one condition: alcohol had to be banned from the town. Known for its rowdy saloons, Mayfield rejected his requests for reform. This led Stanford to form a Temperance Town which he named Palo Alto. In 1894 Stanford purchased 740 acres from Timothy Hopkins (son of Railroad Barron, Mark Hopkins) to accommodate his new town and train station.

Postcard of El Palo Alto.

Menlo Park

The original Menlo Park gates as seen here around the turn of the century. Menlo Park Historical Association.

In 1854, two Irishmen (whose wives were sisters) purchased a large plot of land on the San Francisco Peninsula. They built two houses accessed through a single entrance. The men named this new home after their homeland of Menlough near Galway, Ireland. Their common drive was marked by a tall arched gateway marked with the words "MENLO PARK" and the date, August 1854. The Menlo Park gates stood on the west side of El Camino Real, about 500 feet from Santa Cruz Avenue. When the railroad came through in 1863, the station did not yet have a name. The words Menlo Park were visible from the station and it was then adopted as the area's official name. This stop is now the oldest train station in continuous operation in California and has become a State Landmark. The gates were destroyed by a motorist in 1922.

Atherton

Linden Towers was a mansion built by James Flood and was the largest house west of the Rocky Mountains. It featured 43 rooms, a 40 person dining room table and a tower which stretched up almost seven stories. Flood, a self made millionaire, grew up in the New York slums. He made his fortune in the silver mines of Nevada's Comstock Lode. By 1873 he was making $250,000 a month and soon became known as a Silver King.
Bancroft Library.

In 1860 shipping magnate Faxon Atherton purchased 640 acres in what was then known as Fair Oaks. His home, named Valparaiso Park, was located approximately where the Menlo Circus Club is today. Wealthy San Franciscans began to take advantage of the train and established summer homes in the area. Early prominent land owners included one time San Franciscan mayor Thomas Selby who named his 420 acre country estate Almendral. As profiled above, James Flood also built an extravagant mansion on 600 acres where the Lindenwood neighborhood is today. In 1923 the town incorporated at which point they learned that the name Fair Oaks was already taken. The residents settled on the name Atherton in honor of its first prominent resident.

Redwood City

Downtown Redwood City at the corner of Broadway and Main during the 4th of July parade in 1885. Archives Committee of the Redwood City Public Library

Redwood City got its start as a port town during the Gold Rush. The neighborhood around the port soon sprouted hotels, saloons, stables, leather tanners, blacksmiths and ship-making operations to accommodate the influx of people and goods. Initially called Mezesville, the early residents did not care for that name nor did they care for the other unofficial town name of Embarcadero, which means pier in Spanish. Locals took to calling the place Redwood because of the large amount of redwood lumber coming down from the foothills and shipped from the port up to San Francisco. In 1856, the county of San Mateo was formed and Redwood City was adopted as the official town name and appointed the County Seat of San Mateo County.

Woodside

The Woodside store was built in the early 1850s to provide food, alcohol, post office services and as the local bank (being the only place in the area with a safe) to serve the lumbermen in the area. San Mateo County Historical Association Collection

The first English-speaking settlers arrived in what would become Woodside in the mid 1830s to log the vast redwood forests. In 1849, 20 year-old Matthias Parkhurst purchased 127 acres of land in the area and named it Woodside. Parkhurst partnered with 33 year-old dentist Robert Tripp to build a market on Kings Mountain Road, which is now preserved as a San Mateo County museum (pictured above). By 1855, there were at least eight sawmills and one shingle mill which was the height of logging operations in Woodside. Gradually, the sawmills were replaced by farms, orchards, ranches, and vineyards. In the 1880's wealthy San Francisco families began to build country estates in Woodside. The Pioneer Hotel was built in 1882, and the original facade still stands. Woodside was incorporated in 1956 to prevent urbanization and it retains a rural residential feel to this day.

Portola Valley

Alpine Inn Beer Garden (formerly Rossotti's) located on Alpine Road circa 1940. This tavern is one of the oldest in the state, dating back to 1852 when it was constructed by Felix Buelna, a former mayor of San Jose. It originally served as a clubhouse for Mexican American Californios. Over the years, the roadhouse benefited greatly from Stanford University's dry campus policy. The drinking establishment was just far enough away to prohibit university officials exerting control over what was served. Silicon Valley Historical Association.

Portola Valley was named for Spanish explorer Gaspar de Portolá, who led the first party of Europeans to explore the San Francisco Peninsula in 1769. Around the turn of the century Andrew Hallidie—the inventor of San Francisco's cable cars—lived on a large parcel of land he named Eagle Home Farm. His estate extended from Portola Road to Skyline Boulevard. To traverse his property and to showcase his technology, he built a 7,300 foot long tram connecting his home in Portola Valley to the ridge of the Santa Cruz Mountains. The trip took approximately 30 minutes to the top. Hallidie donated a portion of his land as a site for a school and a small village developed around it on Portola Road. The town was officially incorporated in 1964.

Searsville

On the left is the Searsville Exchange, on the right is the Eikerenkotter Hotel. Wikimedia

Between 1850 and 1890, there was a town by the name of Searsville which stood along Sand Hill Road. The small lumberjack settlement consisted of around a hundred residents and featured a blacksmith shop, post office, school, multiple saloons, a store called the Searsville Exchange and a hotel as seen in the photo above. As logging operations in the area started to die down, the town was largely vacated. In 1887 the land was sold to Spring Valley Water Works. The water company built a dam which created the Searsville Reservoir which still exists today. There are conflicting stories as to how the town was vacated and legends evolved to tell of the of Searsville ghosts at the bottom of the lake. Stanford University acquired the water company in 1919 and for a time the lake was used as a local swimming hole. Today the reservoir is part of the Jasper Ridge Biological Preserve which is overseen by the university.

MILPITAS

Looking up Main Street around 1890. This road was the main road connecting San Jose with Oakland and was the East Bay's equivalent of El Camino Real on the Peninsula. The two-story building on the left corner was known as French's Hotel, later renamed Milpitas Hotel. Milpitas Historical Society.

The land in modern-day Milpitas was originally part of a few different ranchos: Rancho Rincon de Los Esteros (Spanish for the corner of the wetlands) Rancho Milpitas (frequently translated as little corn fields) and Rancho Los Tularcitos (Spanish for little tule marshes). There was a period of rapid immigration between the 1850-1890. Main Street (known at the time as Mission Road) was the heart of town and was home to a hotel, post office and a number of saloons. For over a century, Milpitas served as a popular rest stop for travelers on the old Oakland–San Jose Highway. In 1954 the area was concerned that they would soon be swallowed up by a rapidly expanding San Jose. At that point, Milpitas residents decided to incorporate.

FREMONT

Mission San José was constructed in 1797. Although the original adobe structure was destroyed in the earthquake of 1868, he church has been reconstructed and is considered to be a near-perfect replica. Wikimedia.

Located in present day Fremont, Mission San José was founded in 1797 and was known to be one of California's most prosperous missions. By 1830, there was a population of almost 2,000 people, 12,000 cattle, 13,000 horses and 12,000 sheep. The animals would roam between present day San Jose and Oakland. When Mexico declared their Independence in 1834, the mission fell into a period of decline. With the discovery of gold in California, a small town developed around the mission and flourished with an influx of inhabitants. The original mission building was converted into a general store, saloon and hotel. From 1912-1915 the Niles section of Fremont was the home of California's then nascent motion picture industry. Charlie Chaplin famously filmed several movies in the area. In 1956, the towns of Niles, Mission San Jose, Centerville, Irvington, and Warm Springs banded together and incorporated as the City of Fremont in 1956.

Charlie Chapman filming The Tramp in 1915 on Niles Canyon Road.

TRADE MARK
PACIFIC PACKING CO.
STANDARD BRAND
WHEN OPEN, REMOVE THE FRUIT FROM THE CAN.
Embarcadero
BRAND U.S. NO. 1
Fancy
SANTA CLARA VALLEY
NET WEIGHT 46 LBS.
PRODUCE OF U.S.A.
PEARS
POPPY
BRAND
CALIFORNIA
PEARS
PACKED AND SHIPPED BY
DAY & YOUNG
LUXURY
BRAND
ROYAL ANNE CHERRIES
NET WEIGHT 1 LB. 14 OZ.
NET WEIGHT 25 LBS.
Sunsweet
Sunsweet
CALIFORNIA'S
NATURE-FLAVORED
APRICOTS
PREPARED WITH SULPHUR DIOXIDE
GROWN AND PACKED BY
CALIFORNIA PRUNE AND APRICOT GROWERS INC.
MAIN OFFICE: SAN JOSE, CALIFORNIA.
PACIFIC PACKING CO
SAN JOSE
CAL. U.S.A.
Nº 2½
EXTRA
QUALITY.
APRICOTS
MOUNT HAM
FRESH
WHIT
LICK OBSERVATORY
MOUNT HAMILTON
SANTA CLARA COUNTY
CALIFORNIA
McCURDY
EXTRA FANCY SANTA CLARA VALLEY
U.S. Nº1
CALIFORNIA
PEARS
GROWN AND PACKED BY
SWEETBRIER ORCHARDS
SAN JOSE, CALIF., U.S.A.
BEST CHOICE
BRAND
CALIFORNIA PEARS
SANTA CLARA VALLEY FRUIT & PRODUCE CO.
GROWERS AND SHIPPERS
SAN JOSE, CALIFORNIA
ELK
FANCY
FR
PACKED BY
JOHNSON
CALIFORNIA PACKING CO.
Yellow Peaches
CALIFORNIA
EUREKA
QUALITY GUARANTEED
NATURAL FLAVOR RETAINED.
STANDARD
PACKING CO.
Packed at SAN JOSE, Santa Clara County, CALIFORNIA.
PACH
YELLOW FREE

Farm land in Santa Clara Valley in the 1930s. Silicon Valley Historical Association.

Towards the end of the 1800s, farmers realized that fruit and nuts grew very well in the temperate valley. Because the return on investment in an orchard was much higher than a grain field, land values started to rise. According to the book published in 1896 entitled *Sunshine, Fruit and Flowers,* land that once produced grain and cost $15 per acre, now commanded $200–$600 per acre for the most suitable plots for fruit orchards. While everything from cherries to pears, berries, almonds, apricots, apples and tomatoes fared extremely well in the valley, it was the prune that was the most profitable, and therefore the most extensive crop. Around this time, over 200 million pounds of fresh, dried and canned fruit was exported from the valley every year. There was no better place on earth to grow fruit than Santa Clara Valley, and soon the area became known as "The Valley of Heart's Delight."

Four: Valley of Heart's Delight

Orchards

Apricot Drying Racks in the Valley circa 1936. Silicon Valley Historical Association.

In the 1870s wheat farming in Santa Clara County became uneconomical due to soil degradation and new county property tax laws. Small fruit orchards replaced the large wheat farms and farmers branched out into a wide array of crops: prunes, apricots, peaches, pears, cherries, oranges, almonds, walnuts.

Farm land in Santa Clara Valley in the 1930s. *Silicon Valley Historical Association.*

In 1919 there were nearly 100,000 acres of fruit trees, 3,000 acres of wine grapes, and 85,000 acres of grains, vegetables and berries. In that same year, the farmers in Santa Clara County received $45 million from the products of their trees (equivalent to $590 million in today's dollars). The crop was by far the largest ever raised in the Santa Clara Valley.

FRUIT PACKING, CANNING & DRYING

A fruit processing center in the 1950s. Silicon Valley Historical Association.

In 1871 the first cannery, box and fruit packing companies opened in Santa Clara Valley. The development of food processing technology, including canning and drying, created a new opportunity for growers in Santa Clara Valley. During the heyday, the canneries in Santa Clara Valley shipped ten million cans of fruit and vegetables worldwide annually. During the canning season, the air would be filled with the fragrance of cherries, plums, tomatoes, and other fruits being processed. In 1917 when the U.S. entered World War I, the canneries enjoyed an increased demand for canned fruit to feed the soldiers. In addition, the lack of a male work force due to the draft created new opportunities for women to enter the workforce.

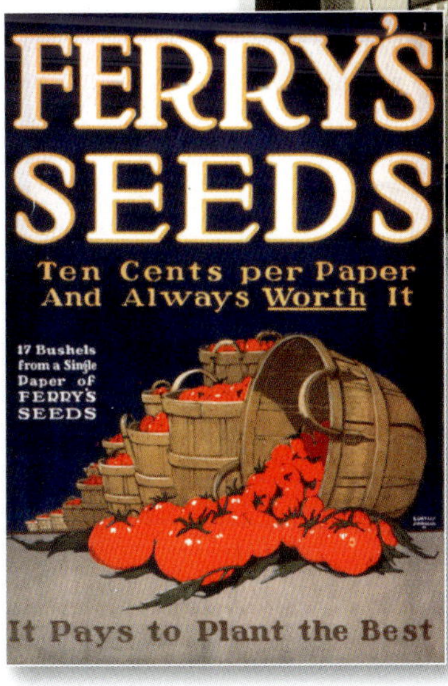

Workers on an assembly line at the Ferry-Morse Seed Co. in Mountain View in 1960. City of Mountain View.

Ferry–Morse Seed Co. started in Mountain View in the 1870s and grew to be the largest seed producer in the world. At harvest time, the company employed 500 people. Ferry-Morse Seed Co. was located in a triangular parcel of land bounded by Mountain View–Alviso Road (now Highway 237), Whisman Road and Evelyn Avenue. In 1985 the plant closed and moved to Modesto and shortly thereafter Hewlett Packard moved into the former Ferry–Morse site.

Fruit packing, canning & drying

Sunsweet Packing Plant in Campbell.

By 1960 there were 215 fruit processing operations in the Valley. Not only was fruit processing the largest employer, it provided a market for many other local industries including: label makers, fruit box makers, can makers, machinery providers and shippers. Restaurants and bars also popped up near the processing plants to cater to the workers.

Libby Cannery on the railway in Sunnyvale - date unknown. Sunnyvale Historical Society.

Libby Cannery opened its doors in Sunnyvale in 1909. At the time, workers received about $1.50–$3.75 per day at piece rate; in today's dollars that would be a daily wage of $40–$100. By 1922 Libby's was the largest cannery in the world. Like the Libby Cannery noted here, most canneries and fruit packers located themselves along the rail line for easy transportation of their canned goods.

Blossom tours

Postcard of a field of sweet peas near San Jose, circa 1908. California History Center.

Postcard of blooming orchards in Los Gatos, circa 1911. California History Center.

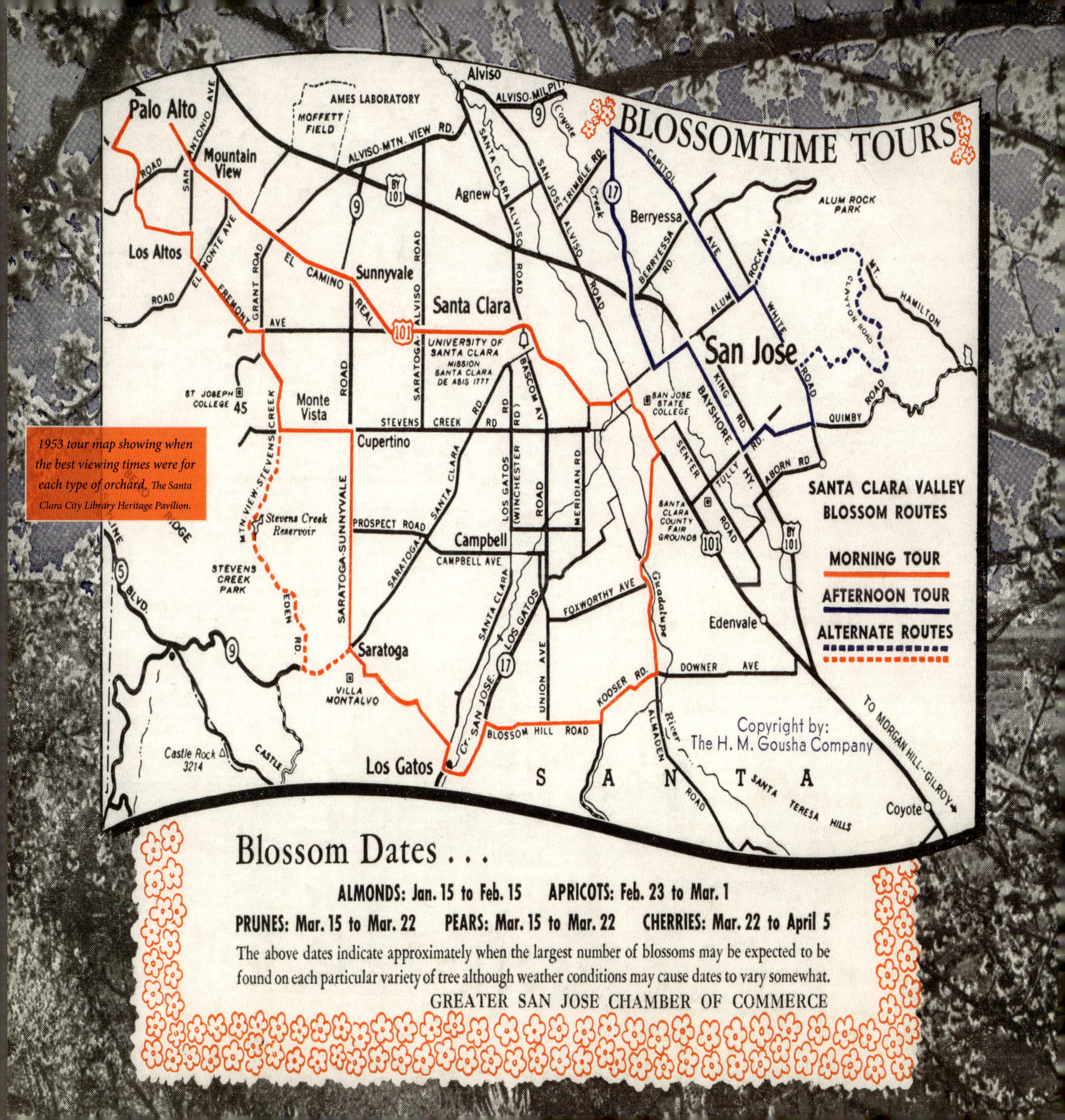

1953 tour map showing when the best viewing times were for each type of orchard. The Santa Clara City Library Heritage Pavilion.

South Pacific Coast Railway in the Santa Cruz Mountains, circa 1892. San Lorenzo Valley Museum.

In order to get from San Francisco to San Jose prior to 1850, the only mode of transportation was via stagecoach using an old French omnibus pulled by a mule. The trip took nine hours and cost two ounces of gold or $32.00. After discovery of gold, the large influx of immigrants and the economic vibrancy of the area demanded more modern modes of transportation. Two railways were built, the first from San Francisco to San Jose, with a later branch covering the West Valley via Los Altos. The second railroad was a narrow gage railway and stretched from Santa Cruz to Alviso. Around the turn of the turn of the twentieth century the Valley floor was covered with streetcar lines. The ports of Alviso and Redwood City were major hubs of commerce as ships would travel up and down the bay carrying passengers and freight. The beginning of the 20th century saw the introduction of the personal automobile. One of the most visible changes resulting from the personal automobile was the replacement of train tracks with freeways. For example, in 1959 the railway tracks that once served the Santa Cruz mountains were torn out to make way for Highway 17. In 1962 Foothill Expressway was built upon the right of way that previously served as the tracks for the Los Altos branch of the Peninsular Railway. The buildings at Loyola Corners in South Los Altos are historical railroad station buildings. In 1949 the first commercial air flight landed in San Jose which ushered in the era of air transportation for Valley residents.

Five: Infrastructure—Planes, Trains & Automobiles

The San Francisco & San Jose Railroad

After California gained statehood in 1850, there was increased need for travel between San Francisco and San Jose. The San Francisco and San Jose Railroad was built between 1861 and 1864. Upon completion, the line was 49.5 miles long and the trip took 2 hours and 20 minutes. By 1896 there were 15 daily trains from San Jose to San Francisco, which cost the rider $1.25 each way. The San Francisco and San Jose Railroad was acquired by The Southern Pacific Railroad in 1868.

Santa Clara railroad station, at the corner of Benton Street and Railroad Avenue, circa 1900. Santa Clara City Library Heritage Pavilion

San Francisco and San Jose Railroad

SAN FRANCISCO AND SAN JOSE RAILROAD.

TIME TABLE

For the government and information of employes only. It is not intended for the information of the public, the Company reserving the right to vary therefrom as circumstances may require.

TAKES EFFECT AUGUST 1ST, 1866.

NORTHWARD TRAINS.							Distance from San Jose.	NAMES OF STATIONS AND PASSING PLACES.	Distance from San Francisco.	SOUTHWARD TRAINS.						
SUNDAYS.			WEEK DAYS.							WEEK DAYS.				SUNDAYS.		
EVE.	EVE.	MOR.	FRT.	EVE.	MOR.	MOR.				MOR.	EVE.	EVE.	FRT.	MOR.	MOR.	EVE.
9.50	6.10	10.30	11.30	6.10	9.30	8.30	49¾	Ar. San Francisco........Dep.		8.20	4.30	5.30	12.30	8.30	9.50	4.10
9.45	6.05	10.25		6.05	9.25	**8.25**	46¼	" * Mission................ "	3½	**8.25**	4.35	5.35		8.35	9.55	4.15
9.40	6.00	10.20		6.00	9.20	8.20	45¼	" * Bernal,................ "	4½	8.30	4.40	5.40		8.40	10.00	4.20
9.30	5.50	**10.10**	10.50	**5.50**	9.10	8.10	43¼	" * San Miguel,.......... "	6½	8.40	4.50	**5.50**		8.50	**10.10**	4.30
9.23	5.43	10.03		5.43	9.03	8.03	40¾	" * School House....... "	9	8.47	4.57	5.57		8.57	10.17	4.37
9.16	5.36	9.56		5.36	**8.56**	7.56	38¼	" * 12 Mile Farm,...... "	11½	**8.56**	5.06	6.06		9.06	10.26	4.46
9.09	5.29	9.49		5.29	8.49	7.49	35¼	" * San Bruno,........... "	14½	9.01	5.11	6.11		9.11	10.31	4.51
9.00	5.20	9.40		**5.20**	8.40	7.40	33	" Millbrae,.............. "	16¾	9.10	**5.20**	6.20	2.00	9.20	10.40	5.00
8.50	**5.10**	**9.30**	9.25	5.10	8.30	7.30	29	" San Mateo,............ "	20¾	9.20	5.30	6.30	2.20	**9.30**	10.50	**5.10**
8.37	4.57	9.17	9.10	4.57	8.17	7.17	24¾	" Belmont,............... "	25	9.33	5.43	6.43	2.35	9.43	11.03	5.23
8.30	4.50	9.10	8.50	4.50	8.10	7.10	21¼	" Redwood City,....... "	28½	9.40	5.50	6.50	2.50	9.50	11.10	5.30
8.20	4.40	9.00	8.20	4.40	8.00	6.55	17¼	" Menlo Park,.......... "	32½	9.50	6.00	7.00	3.15	10.00	11.20	5.40
8.13	4.33	8.53	8.10	4.33	7.53	6.50	15¼	" Mayfield,............... "	34½	9.57	6.07	7.10	3.30	10.07	11.27	5.47
8.02	4.22	8.42	7.45	4.22	7.42		9¾	" Mountain View,..... "	40	10.08	6.18		4.00	10.18	11.38	5.58
7.48	4.08	8.28	7.27	4.08	7.28		6	" Lawrence's,........... "	43¾	10.22	6.32		4.20	10.32	11.52	6.12
7.40	4.00	8.20	7.15	4.00	7.20		3¼	" Santa Clara,........... "	46½	10.30	6.40		4.35	10.40	12.00	6.20
7.30	3.50	8.10	7.00	3.50	7.10			Dep. San Jose,................. Ar.	49¾	10.40	6.50		5.00	10.50	12.10	6.30

☞ * Indicates Flag Stations.

☞ 1. The figures set against San Francisco Northward, and San Jose Southward, are the times to REACH those Stations. The other figures represent the times of LEAVING the Stations against which they are placed.

☞ 2. The figures in large black type designate the meeting places of trains, and the time of passing.

☞ 3. In case freight train is not at passing place on time, passenger train will proceed with caution until such train is passed.

☞ 4. Freight train must keep clear of all regular passenger trains at least ten minutes.

☞ 5. All irregular trains must keep clear of regular trains at least fifteen minutes.

☞ 6. Freight train leaving San Jose at 7.00, P. M., will wait the arrival of the 4.30, P. M., train from San Francisco.

☞ 7. The 7.00, P. M., freight from San Jose must not pass Mayfield until the 5.30, P. M., train from San Francisco arrives.

☞ 8. Should freight train leaving San Francisco at 12.30, A. M., be delayed it must keep clear of trains leaving Mayfield at 6.50, A. M., and San Jose at 7.10, A. M.

☞ 9. **All Regular Passenger Trains will wait twenty (20) minutes at passing places designated in the Time Table, and then proceed, always keeping twenty (20) minutes behind Card time until the expected train is passed. When both trains lose the twenty (20) minutes, they will proceed with flag in advance, and run with caution until trains pass.** See General Rules, No. 24.

☞ 10. **San Francisco and San Jose are terminal stations. All regular passenger trains will leave terminal stations on time, notwithstanding trains may be due. A train loses its right of road which does not reach the terminus before the starting time of a regular train. No excuse will be received for neglect of any of these rules.**

CHAS. B. GOULD, Acting Superintendent.

Oct.31.st.192

1923

Railroad time table from 1866. South Bay Historical Railroad Society.

Menlo Park Train Station, looking towards Palo Alto, 1885. Palo Alto Historical Association.

The train leaving the Palo Alto station and steaming towards the famous El Palo Alto tree in 1894. Palo Alto Historical Association.

San Francisco and San Jose Railroad

Postcard of the Southern Pacific Railroad station in San Jose, 1911. California Room.

In 1908 Southern Pacific Railroad began running steam train service through Los Altos with 5 trains per day. This was a secondary Southern Pacific route that passed through the Stanford University campus in Palo Alto. It was abandoned in 1962 to make way for the Foothill Expressway. Image taken in Palo Alto. Los Altos Historical Society.

Murphy Station railroad stop in Sunnyvale, looking east circa 1911. Sunnyvale Historical Society.

The Pacific Coast Railway

A restored train from the early 1900s currently serves tourists visiting the Santa Cruz mountains.

Originally known as The Santa Clara Valley Railroad, the South Pacific Coast Railway was a narrow gauge railway that was constructed to serve the lumbermen and fruit growers as shipping prices became cost prohibitive on the San Francisco–San Jose Railroad. It was built in two sections: the first from the port in Newark to Los Gatos, with stops in Santa Clara and Alviso; and the second from Los Gatos to Santa Cruz. The full train track was 45 miles long and opened in 1880. The invention of the refrigerated railroad car in the late 1800s allowed the Valley to become the largest fruit production and packing region in the world as this allowed them to ship large quantities of perishable goods and keep them fresh during transport.

Loading a refrigerated train cabin, 1890.

Silicon Valley Historical Association.

The Pacific Coast Railway, steaming past the Santa Cruz Beach Boardwalk.

Advertisement for Casa del Rey, a casino and resort on the Santa Cruz beach.

While initially set up as a freight service, tourism from San Francisco to the beaches in Santa Cruz became a popular service and special trains were set up to cater to the weekend revelers. This train service was called the Suntan Special and a round trip ticket cost $1.25. Resorts sprung up throughout the mountains and along the beach in Santa Cruz to cater to the new influx of tourists.

Postcard of the Santa Cruz Beach Boardwalk from its heyday.

The Alviso Regatta

Before train service, the most efficient way to get from the Valley to San Francisco was to take a boat from Alviso, which cost $35. After passengers migrated to the San Francisco–San Jose Railway in 1864, Alviso remained a popular trading post for merchants. This small port town became even more important after the introduction of the Pacific Coast Railway in 1880 that brought lumber and fruit by rail to Alviso to be placed on a steamship bound for San Francisco. The narrow gage railway that served the port of Alviso was abandoned in 1942. In 1973, nine acres containing 11 historic structures were added to the National Register of Historic Places.

Steamboat landing in Alviso, 1895. San Jose History Center.

The Alviso Regatta in 1912. The South Bay Yacht Club in the background was built in 1903 and still serves as the club's headquarters today. California Room.

The Port of Redwood City

The Port of Redwood City has been operational since 1851. In the beginning, the port mainly served as a transportation hub for the redwood trees that had been harvested in the foothills, hauled down Woodside Road and loaded onto ships to be transported to San Francisco. The port is located 18 nautical miles south of San Francisco and is the only deep-water port in the South San Francisco Bay. Today, the Port of Redwood City is still active and continues to be remembered as the birthplace of ship-building on the North American West Coast.

Port of Redwood City, circa 1870. San Mateo County Historical Association Collection

Interurban Transportation

1907 Streetcar in Los Gatos. Santa Clara City Library Heritage Pavilion.

1917 timetable. California History Center.

1915 Streetcar on Santa Cruz Avenue in Los Gatos. California History Center.

1937 streetcar traveling eastbound on The Alameda at Newhall Street in San Jose.

San Jose was the first city in California to have a streetcar, which was introduced in 1890. Trolley lines eventually covered 70 miles of track and many lines provided train service every seven minutes between the downtown San Jose core and then outlying towns including Santa Clara, Monta Vista, Los Altos, Los Gatos, Alum Rock, and Beryessa, as well as further out to Palo Alto and Stanford. In 1920 streetcar trains reached the center of San Jose from the far points of the line in about 15–20 minutes. One could board a car in say, Saratoga, and travel to Stanford for a football game or a day of shopping and appointments in downtown San Jose, as service was every half-hour. It was a true modern metro network. Around this time, there were more than 30,000 streetcars in the United States, using approximately 15,000 miles of track and carrying more than 11 billion passengers a year. These electric streetcars ran on tracks and used a pole to draw power from an overhead wire. Yet only a few decades later these streetcars were abandoned; the last streetcar service in San Jose was in 1938.

Introduction of the Personal Automobile

A horseless carriage circa 1910 in Palo Alto. Silicon Valley Historical Association.

The demise of the rail system was due to the proliferation of the personal automobile. It was the large-scale assembly line manufacturing of affordable automobiles by Henry Ford in the early 1900s that put cars in the hands of average citizens. Ford's cars came off the line in fifteen minute intervals, increasing productivity 800% while using less manpower. His process was so successful that by 1914 an assembly line worker could buy a Model T with four month's pay.

Wine delivery via car, 1920. City of Mountain View.

The traffic jam begins! El Camino Real in 1940. Hoover Tower is in the background. Palo Alto Historical Association

We Paved Paradise & Put Up A Parking Lot

The replacement of streetcars with buses across the country is known as The Great American Streetcar Scandal. General Motors, Firestone Tire, Standard Oil of California, Phillips Petroleum, Mack Trucks, and the Federal Engineering Corporation have been accused of buying over 100 electric surface-traction systems in 45 cities across the country and converting them into bus operation to further their corporate interests and push the US into automobile dependency. The story has been explored several times in print, film and other media, notably in the movies Who Framed Roger Rabbit, Taken for a Ride and The End of Suburbia.

Paving of Highway 9 in 1915. The Cupertino Store is at the right. California History Center.

The Age of the Freeway

Highway 17 in 1960. Ruins of Forbes Mill can be seen on the right. The California Room.

Once cars became the main mode of transportation, the era of the freeway commenced. Highway 101 was commissioned in 1926 as one of the original US highways extending from the Mexican border to Washington State. The portion of the freeway between San Jose and San Francisco was a four lane road with no median and was known as Bloody Bayshore due to the high frequency of accidents. Between 1949 and 1962 the road was upgraded to between six and ten lanes. Interstate 280 was added to the Interstate Highway System in 1955 as an additional route between San Jose and San Francisco. However, the freeway was not completed until the 1970s. In 1959 Highway 17 opened up between Los Gatos and San Jose roughly following the former Pacific Coast Railway.

Highway 101 in 1940.

Highway 280 in 1970. Silicon Valley Historical Association.

San Jose Airport

"Airport Day" -The 1945 inaugural flight into San Jose drew crowds to witness the landing along with the newly constructed runway and terminal building. San Jose Airport Authority.

The San Jose International Airport got its start when San Jose had a population of less than 68,000 and its economy was linked to the thousands of acres of orchards that stretched out from its city limits. Aviation enthusiasm had been growing since the early 1920s when World War I fly boys returned to put on exhibitions and conduct flying schools. In 1945 Jim Nissen leased 16 acres of land in San Jose for a small aviation business, California Aviation Activities, and built the first dirt runway. In 1946 the city approved development of the San Jose Municipal Airport and it opened in 1949.

Jim Nissen and a friend in 1950. Nissen was the grandfather of the San Jose Airport. In 1945 he opened "California Aviation Activities" and built the first dirt runway where the airport stands today. San Jose Airport Authority.

The original San Jose Municipal Airport Staff posed in front of the first terminal building in 1948. San Jose Airport Authority.

San Jose Airport

Many early flights out of San Jose were to transport the Valley's vast agriculture industry. San Jose Airport Authority

The first commercial airline flight that landed at SJC was a DC-3 with two pilots, seven passengers and 2,550 baby chickens. Only the chickens got off at San Jose; the human passengers continued to Los Angeles.

Aerial of the San Jose airport circa 1945. San Jose Airport Authority

Palm Drive into Stanford University, 1891. *Stanford University Archives.*

Six: Stanford University

The Stanford Family

Stanford farm home in Palo Alto circa 1887. Stanford University Archives.

Leland Stanford, who grew up and studied law in New York, moved west to chase the gold rush and ended up making his fortune in the railroad that followed. He was a leader of the Republican Party, Governor of California and later a U.S. Senator. He and Jane had one son who died of typhoid fever in 1884 when the family was traveling in Italy. Leland Jr. was just 15. Within weeks of his death, the Stanfords decided that, because they no longer could do anything for their own child, "The children of California shall be our children." They quickly set about to find a lasting way to memorialize their beloved son.

Stanford family photo taken in 1881. Stanford University Archives.

The Harvard of the West

Middlefield Road in 1890. Silicon Valley Historical Association.

Stanford University Postcard. Stanford University Archives.

Leland Stanford acquired more than 8,000 acres of adjoining properties to house his new university. The Stanfords engaged Frederick Law Olmsted, the famed landscape architect who created New York's Central Park, to design the physical plan for the university. On October 1, 1891 Stanford University opened its doors after six years of planning and building. The first student body consisted of 555 men and women, which was quite progressive at the time. The original faculty of 15 was expanded to 49 for the second year.

Stanford students in the 1920s. Stanford University Archives.

Stanford University's first president, David Starr Jordan circa 1880. Wikimedia

The legacy of Stanford University's first president in a complicated one. David Starr Jordan grew up on a farm in upstate New York and in an unorthodox move, attended a local girls' high school. He was part of the pioneer class of undergraduates at Cornell University and went on to become the president of Indiana University. He was offered the opportunity to become the inaugural president of Stanford University based on his vision of a non-sectarian, co-educational school with a liberal arts curriculum. On the one hand: He helped fight for woman's rights including the right to vote. He was a pacifist. He was an early environmentalist, serving as president of the Sierra Club and founding the Sempervirens club in an effort to preserve the redwood forests. But on the other hand, he was a firm believer in eugenics, a view which he proselytized throughout his life. In 2018, Palo Alto made the decision to rename Jordan Middle School in a move to stop paying homage to his legacy.

Herbert Hoover

Herbert Hoover campaigning for President in Palo Alto in 1928. Stanford University Archives.

Hoover Tower in 1941. Stanford University Archives.

Herbert Hoover entered Stanford University in 1891, its inaugural year, after flunking all the entrance exams (except mathematics) and then being tutored for the summer in Palo Alto. Hoover claimed to be the very first student at Stanford, by virtue of being the first person in the first class to sleep in the dormitory. While at the university, he was a part of the inaugural Big Game versus rival University of California (Stanford won). Hoover Tower, inspired by the cathedral tower at Salamanca, was built to commemorate Stanford's 50th anniversary. It was finished in 1941, stands 285 feet tall and houses the Hoover Institution Library and Archives. Hoover had amassed a large collection of materials related to early 20th century history and donated them to Stanford prior to becoming president.

John F. Kennedy

John F. Kennedy circa 1961.

John F. Kennedy attended Stanford University in 1940 as a graduate student for a semester before leaving to enlist in the Navy. In a letter to a friend he wrote, "*Have become very fond of Stanford. Everyone is friendly. The gals are quite attractive—it's a very good life.*"

L'Omelette was a cozy, wood-paneled restaurant that looked something like a French farmhouse located in the stretch of El Camino Real then known as Restaurant Row. Legend has it that while at Stanford Business School, JFK could often be found at their bar.

Palo Alto Historical Society.

Fred Terman: The Father of Silicon Valley

Fred Terman is not given enough credit for his seminal role in creating what the Valley is today. As an engineering professor at Stanford University, he encouraged his students to start their own companies which resulted in the formation of companies like Hewlett Packard and Litton Industries. He is also part of the team credited with the idea of leasing University land to high-tech firms which paved the way for Stanford Research Park. He catapulted Stanford into the ranks of the world's first class educational institutions by chasing research grants from the Department of Defense.

Fred Terman was a Palo Alto native. His father, Lewis Terman, was a Stanford professor and is credited with popularizing the IQ test. As a teenager, Fred Terman became interested in the relatively new field of radio electronics and operated an amateur station in Palo Alto in 1917. Terman completed his undergraduate degree in chemistry and his master's degree in electrical engineering at Stanford University. He went on to earn an ScD in electrical engineering from MIT in 1924, and then returned to Stanford in 1925 to become a member of the engineering faculty. From 1925 to 1941, Terman designed a course of study and research in electronics that focused on work with vacuum tubes, circuits, and instrumentation.

During World War II, Terman directed a staff of more than 850 at the highly secret Radio Research Laboratory at Harvard University. This stealthy organization was the source of radar-blocking jammers and tunable receivers to detect radar signals used during the War. These countermeasures significantly reduced the effectiveness of enemy radar-directed weapons. After the war, Terman returned to Stanford and was appointed Dean of the School of Engineering. Under his leadership, the department received much of its funding from the Military. He served as Provost at Stanford from 1955 to 1965. During his tenure Terman greatly expanded the science, statistics and engineering departments. It was during this period that Professor Terman became concerned with the lack of economic opportunities for Stanford Engineering graduates in the Bay Area. He was part of the team that convinced the University to create the Stanford Industrial Park (now Stanford Research Park) wherein the University leased portions of its land to high-tech firms. Companies such as Varian Associates, Hewlett Packard, Eastman Kodak, General Electric, and Lockheed moved into Stanford Industrial Park. Fred Terman helped turn the Bay Area into a hotbed of innovation, and paved the way for later generations who would create Silicon Valley.

Fred Terman with his famous protégés in 1939. From left to right: Dave Packard, Bill Hewlett and Fred Terman. Stanford University Archives.

Stanford Research Park

Meeting in 1950 in preparation for the construction of Stanford Industrial Park. Silicon Valley Historical Association.

Stanford Industrial Park (now known as Stanford Research Park) was built in 1951 and served as the world's first cutting-edge technology-focused office park. The park covers 700 acres and is bounded by Page Mill Road and El Camino Real. Early tenants included Varian Associates, Hewlett Packard, General Electric, and Lockheed. Currently the park houses 162 buildings which are populated by 140 different companies who employ 23,000 workers.

Aerial taken in 1960 of the early Stanford Industrial Park. Sunnyvale Historical Society.

Stanford Research Park

Taken in 1960, this aerial is centered on Page Mill Road looking east with Junipero Serra Boulevard visible at the bottom. Palo Alto Historical Association.

As a comparison, this aerial was taken in 1985 and is centered on Page Mill Road with the Palo Alto Square office complex on the left. *Palo Alto Historical Association.*

STANFORD RESEARCH PARK IN 1950
EL CAMINO REAL
PAGE MILL
FOOTHILL EXPRESSWAY
ARASTRADERO

STANFORD RESEARCH PARK IN 1982
EL CAMINO REAL
PAGE MILL
FOOTHILL EXPRESSWAY
ARASTRADERO

Companies Founded by Stanford Alumni

DID YOU KNOW...? If companies founded by Stanford graduates formed an independent nation, it would be the 10th largest economy in the world. California alone is home to about 18,000 firms created by Stanford alumni, generating annual worldwide sales of about $1.27 trillion and employing more than three million people. Worldwide, Stanford alumni and faculty have created 5.4 million jobs since the 1930s and have created nearly 40,000 companies. These findings are the product of a study conducted in 2011 by Charles Eesley, an assistant professor in the Stanford School of Engineering, and computer scientist William F. Miller of the Stanford Graduate School of Business. The study, entitled *Stanford University's Economic Impact via Innovation and Entrepreneurship,* analyzed the results of a large survey of Stanford alumni and faculty. Below is a sampling of companies which had at least one founder who either attended Stanford as a student or were faculty.

Instagram

intuit

Rambus

DOLBY

charles SCHWAB

CISCO

NETFLIX

Zillow

Google

Expedia

PayPal

varian

TRADER JOE'S

Sun microsystems

ATARI

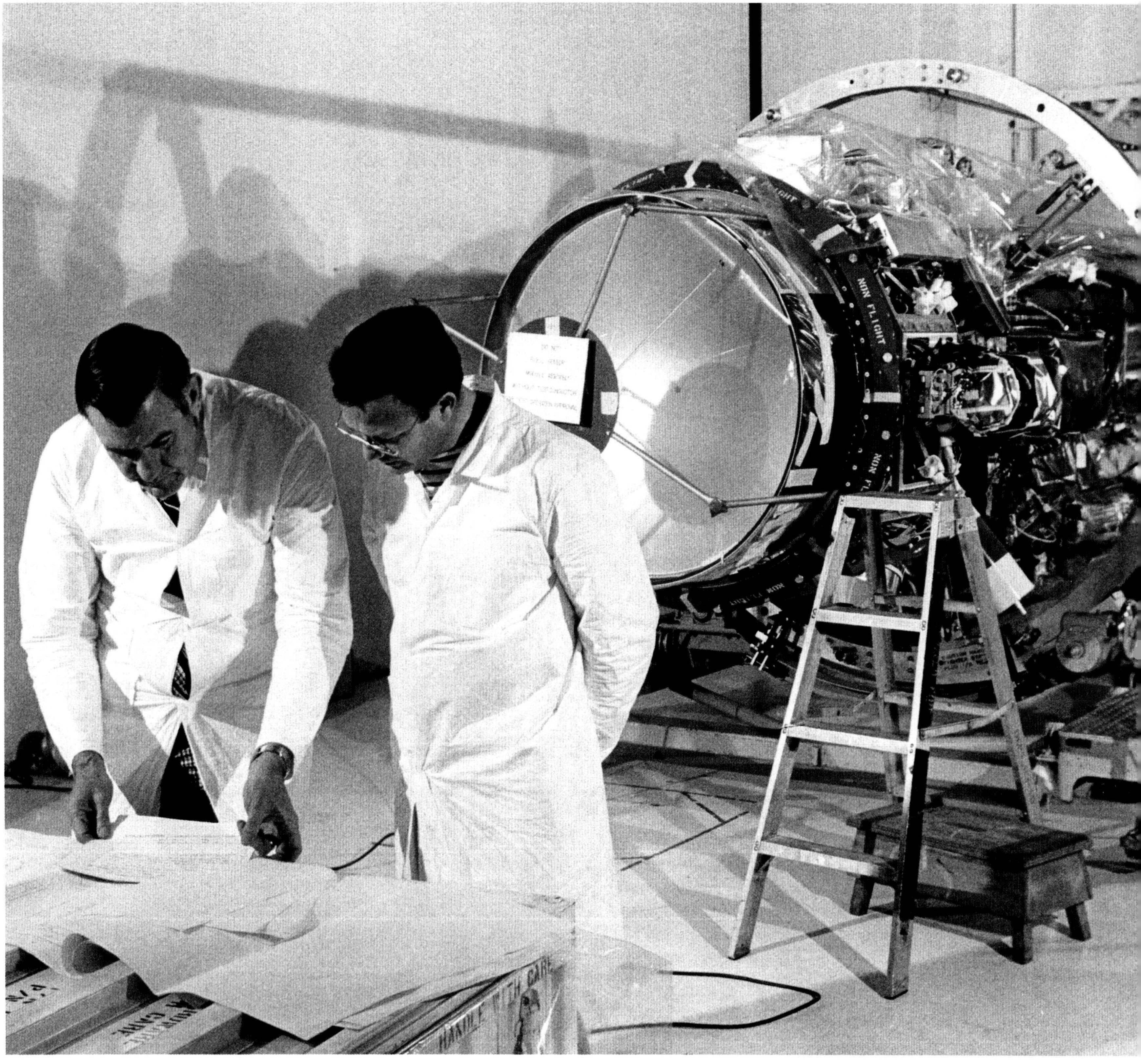

Lockheed engineers working on the Seasat, a satellite designed by Lockheed for NASA. *Lockheed Martin.*

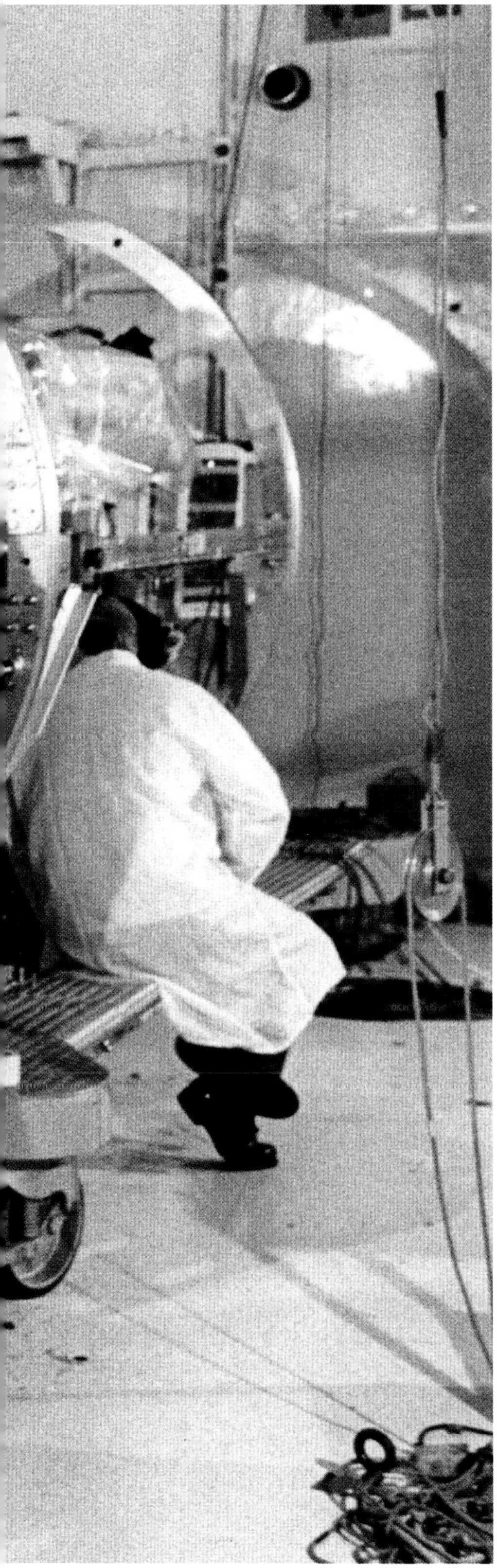

Dating as far back as 1900, the true angel investor of Silicon Valley has been the United States Department of Defense. As new ideas born out of Stanford University began to give way to new industries, Palo Alto became a testing ground for radio and microwave technologies and the area was soon considered a technical beacon. In the beginning, almost all of the funding for these fledgling industries came from the U.S. Government. The money flowed into the Valley from the DOD through both World Wars and the Cold War. It was in Palo Alto where a Stanford Graduate founded the Federal Telegraph Company. FTC is credited with the creation of the radio communication technology which allowed US Naval Ships to communicate with each other for the first time, just in time for World War I. In the 1930s, the residents of Santa Clara Valley pooled their money together to acquire and donate the land for what would become Moffett Field, which turned out to be a sound investment as many thousands of local jobs were created through this acquisition. Also in the 1930s and 1940s two early start-up companies born out of Stanford University, Hewlett Packard and Varian Associates, expanded rapidly thanks to their relationship with the military. During World War II, the U.S. Government and industry invested an estimated $800 million in defense plants in the Bay Area, constructing about 70 new industrial plants in a few short years. During the Cold War and especially after the Soviets launched Sputnik in 1957, the US Government increasingly turned its attention to the potential of computers and aerospace. As weapons grew smarter and reconnaissance became the backbone of the American defense strategy, the Department of Defense focused on smart weapons, smart sensors, and stealth. At the heart of these strategies were microwaves, silicon chips, electronics and computers.

Seven: The Valley's Secret Weapon

The Birth of the Electronic Age

Electrical relays circa 1910. Silicon Valley Historical Association.

As electricity caught on during the first part of the 1900s, the Bay Area (as opposed to the East Coast) was faced with the problem of carrying high tension voltage over long distances. Stanford professors and students in the newly created electrical engineering department helped solve the problem. They inaugurated a cooperative model between university and industry—the Bay Area's electrical power companies employed the Stanford high voltage laboratory and the bright minds who worked there. Together, they were able to develop the technology necessary for long distance electric power. This was the beginning of the revolutionary partnership between Stanford and private industry that would transform the area into a technological powerhouse.

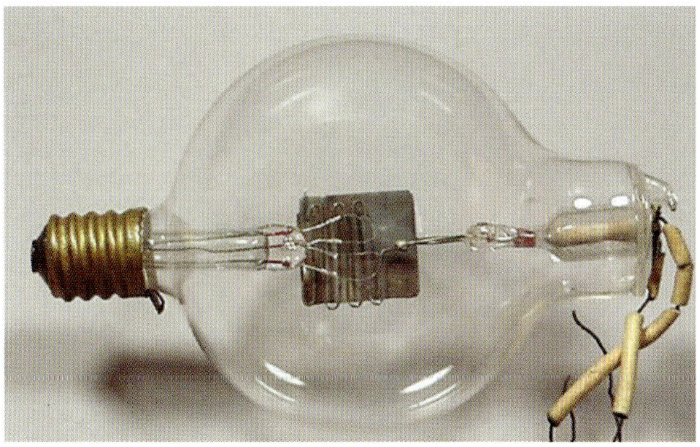

The Audion was a glass tube which was evacuated of most of its gases and contained three electrodes which, when subjected to a small electrical signal, could provide an amplifying effect. The Audion was the first vacuum tube of its kind and was critical to the development of electronic technology.

Lee de Forest with his invention, the Audion, in 1906.

In 1906 an inventor by the name of Lee de Forest invented the Audion which helped usher in the widespread use of electronics. Thus, he is considered a father of the electronic age. Vacuum tubes were critical to the development of electronic technology, which in turn drove the expansion and commercialization of radio broadcasting, television, radar, sound recording, large telephone networks as well as analog and digital computers. Also significant is the fact that Stanford University's President at the time, David Starr Jordan, invested $500 in the development of Lee de Forest's invention, which many say was the first significant venture capital investment in the region.

Federal Telegraph Company

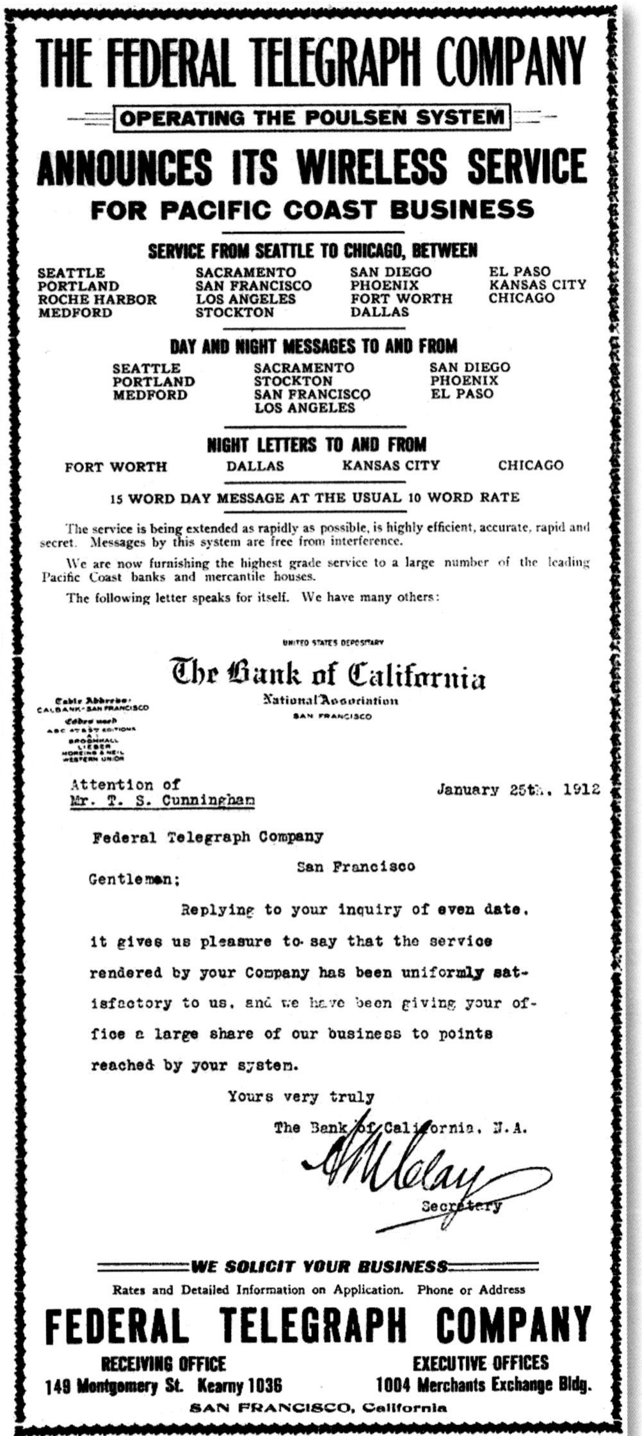

THE FEDERAL TELEGRAPH COMPANY

OPERATING THE POULSEN SYSTEM

ANNOUNCES ITS WIRELESS SERVICE

FOR PACIFIC COAST BUSINESS

SERVICE FROM SEATTLE TO CHICAGO, BETWEEN

SEATTLE	SACRAMENTO	SAN DIEGO	EL PASO
PORTLAND	SAN FRANCISCO	PHOENIX	KANSAS CITY
ROCHE HARBOR	LOS ANGELES	FORT WORTH	CHICAGO
MEDFORD	STOCKTON	DALLAS	

DAY AND NIGHT MESSAGES TO AND FROM

SEATTLE	SACRAMENTO	SAN DIEGO
PORTLAND	STOCKTON	PHOENIX
MEDFORD	SAN FRANCISCO	EL PASO
	LOS ANGELES	

NIGHT LETTERS TO AND FROM

FORT WORTH DALLAS KANSAS CITY CHICAGO

15 WORD DAY MESSAGE AT THE USUAL 10 WORD RATE

The service is being extended as rapidly as possible, is highly efficient, accurate, rapid and secret. Messages by this system are free from interference.

We are now furnishing the highest grade service to a large number of the leading Pacific Coast banks and mercantile houses.

The following letter speaks for itself. We have many others:

UNITED STATES DEPOSITARY

The Bank of California

National Association

SAN FRANCISCO

Cable Address: CALBANK-SAN FRANCISCO

Attention of
Mr. T. S. Cunningham

January 25th, 1912

Federal Telegraph Company
San Francisco

Gentlemen;

Replying to your inquiry of even date, it gives us pleasure to say that the service rendered by your Company has been uniformly satisfactory to us, and we have been giving your office a large share of our business to points reached by your system.

Yours very truly
The Bank of California, N.A.

Secretary

WE SOLICIT YOUR BUSINESS

Rates and Detailed Information on Application. Phone or Address

FEDERAL TELEGRAPH COMPANY

RECEIVING OFFICE
149 Montgomery St. Kearny 1036

EXECUTIVE OFFICES
1004 Merchants Exchange Bldg.

SAN FRANCISCO, California

In 1909 a recent graduate from Stanford, Cyril Elwell, founded the Federal Telegraph Corporation (FTC) in Palo Alto, which created the world's first global radio communication system. Shortly after its incorporation, FTC hired Lee de Forest in order to gain use of the Audion. Stanford's president, David Starr Jordan, was the primary investor and most early employees were Stanford students. Prior to advances made at FTC, long distance transmissions would become faint over long distances. De Forest's vacuum tube enabled repeaters that restored the signal at intermediate points and dramatically improved long distance wireless communications. In 1912 FTC partnered with the Navy and was the pillar of naval communications for the US during WWI. During the war years, employment at FTC rose from 30 to 300.

FTC employees with a 500 kw arc transmitter in background. Palo Alto Historical Association.

The laboratory of Federal Telegraph Company where Lee de Forest worked: 913 Emerson in Palo Alto, which is considered the birthplace of electronics and is now a historic landmark. Palo Alto Historical Association.

Moffett Field

The Macon at Moffett Field, 1934. The Macon was a rigid airship built by the Navy in 1933 to be used for reconnaissance as well as serve as a flying aircraft carrier. It was in service for less than two years. In 1935 the Macon was damaged in a storm and was lost off of California's coast. In order to accommodate such a large aircraft, Hanger One was constructed to be 1138 feet long, 308 feet wide and 198 feet high. Moffett Field Museum.

In an effort to gain the jobs and economic growth born out of government contracts, Santa Clara County residents banded together to purchase 1,000 acres of farmland bordering the San Francisco Bay. The citizens then donated the land to the Navy to be used as the home base for the airship USS Macon. The location proved to be ideal for an airport since the area is often clear while other parts of the San Francisco Bay are covered in fog. The naval air station was authorized by an Act of Congress and signed by President Herbert Hoover in 1931. After the short-lived USS Macon, Moffett Field continued to serve as a military base. After World War II, Moffett Field saw the development and use of land-based anti-submarine warfare and maritime patrol aircraft, mostly coming from a joint venture with neighbor and defense contractor, Lockheed. During the Cold War, daily anti-submarine and maritime reconnaissance missions flew out from Moffett Field to patrol along the Pacific coastline.

Postcard of Moffett Field in 1944. *Moffett Field Museum.*

In 1994, Moffett Field was closed as a military base and supervision of the facilities was turned over to NASA Ames Research Center. In 2015, Google signed a 60 year lease with NASA for the 1,000 acre site which has been valued at $1.16 billion. Google is said to now be using the old hangars as laboratories for cutting-edge technology including robots, drones and Internet-carrying balloons. In 2017, the tech giant began preparations to build a new campus on 42 acres located on the western side of Moffett Field. Known as Bay View, the 1.1 million square foot campus of canopied domes and futuristic offices is slated to be completed by the end of 2019.

Hendy Iron Works

Hendy Iron Works machining floor, 1908. Sunnyvale Historical Society.

Hendy Iron Works was an engineering company that existed from the 1850s to the late 1940s. Originally located in San Francisco, the company relocated to Sunnyvale after being enticed with an offer of free land. The factory is perhaps best remembered for its contribution during World War II. By the end of the war, the Joshua Hendy Iron Works had supplied the engines for 28% of the Liberty ships—more than any other plant in the country. In 1947 the Joshua Hendy Iron Works was sold to the Westinghouse Corporation. In the postwar period the plant continued to produce military equipment including missile launching and control systems for nuclear powered submarines. In 1996 Westinghouse sold the plant to Northrop Grumman, which renamed it Northrop Grumman Marine Systems and continues to operate in the same plant today. The Iron Man museum now operates in a portion of the original building, which details the plant's history.

Hendy built Liberty ship for WWII. Sunnyvale Historical Society.

Hendy Iron Works

The crew responsible for building the Liberty ships poses for a photo to recognize their work, circa 1945. The Iron Man Museum.

President Truman visited the plant in 1943 and is shown here getting a tour of the facility. The Iron Man Museum.

Food Machinery Corporation

The John Bean Spray Pump, the first product of the company that would become FMC. California Room.

Headquarters of Anderson Barngrover in 1920 prior to its merger and name change to Food Machinery Corporation. FMC.

Water Buffalo produced during World War II in FMC's Santa Clara facilities. FMC.

FMC was originally founded in 1883 as the Bean Spray Pump Company in Los Gatos by John Bean. The company's first product was a pump to spray insecticide on the many fruit orchards in the area. In 1928 The Bean Spray Pump Company and Anderson Barngrover merged and after a contest was held for a new name, they decided upon Food Machinery Corporation. By the mid-1930s, FMC was the world's largest manufacturer of machinery and equipment for handling fruits, vegetables, milk, fish and meat products. With the onset of World War II, FMC entered the defense business, making amphibious tractors and tanks for the military. In 1945 the city sold 90 acres of airport land to FMC for expansion when the large employer threatened to leave the county. FMC continues to operate a number of divisions today, although the company no longer has a presence in Santa Clara Valley.

Packard Hewlett or Hewlett Packard?

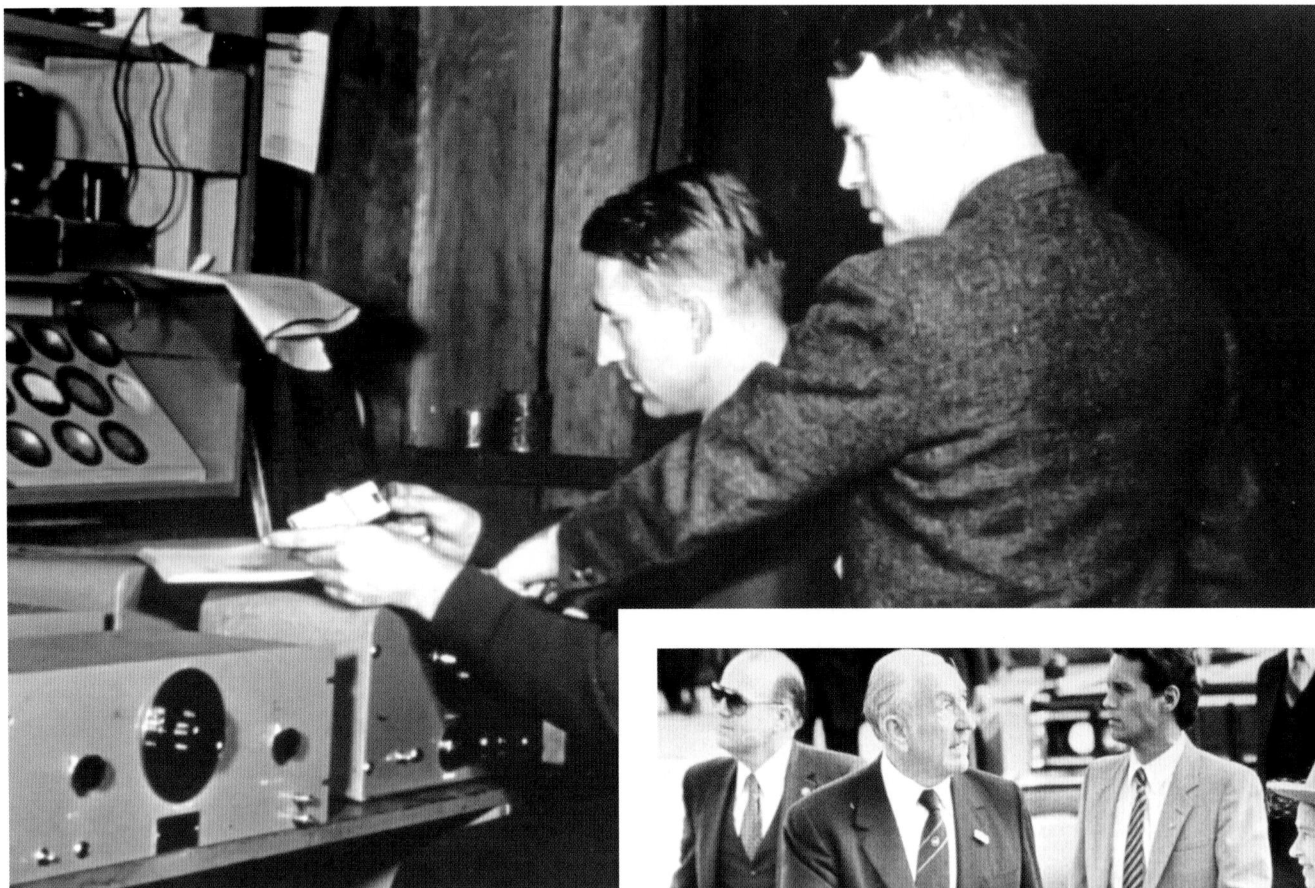

Dave Packard & Bill Hwelett working in the original garage on Addison Avenue in 1939. Palo Alto Historical Association.

Dave Packard and Queen Elizabeth outside HP in 1983. Palo Alto Historical Association.

David and Lucile Packard enjoying a dance together at a company party in 1945. Palo Alto Historical Association.

The two famous couples are shown here thanking their employees at a company party in 1945. Palo Alto Historical Association.

Bill Hewlett and Dave Packard became friends at Stanford in the early 1930s while obtaining degrees in electrical engineering under the mentorship of Fred Terman. After they graduated, Fred Terman convinced them to open up their own company to produce their recent invention, the audio-oscillator. Walt Disney was their first customer who purchased their oscillators for the animation film *Fantasia*. When they first opened for business in the now famous garage, they flipped a coin to determine the name of the company—Bill won. Over the course of their 70 year friendship, there is no record of the two ever fighting. They built their company based on trust. Trust in each other and trust in their employees and customers. This philosophy is now known across the world as 'The HP Way'. Dave was considered the philosopher of the HP Way and Bill was its soul and the man who implemented it every day. This philosophy continues to be the basis by which many start-ups base their corporate model.

A Few of HP's Offices Over The Years

367 Addison Avenue, Palo Alto where work in the now famous garage began in 1939. Palo Alto Historical Association.

395 Page Mill Drive, Palo Alto. In 1942 HP had a total of eight employees and purchased their first building. They designed it to feature an open floor plan for versatility and collaboration, which has become the model for many Silicon Valley start-ups. Palo Alto Historical Association.

1501 Page Mill Road, Palo Alto. In 1957 HP went public and designed their new headquarters to further implement a culture of open communication, teamwork, and innovation.
Palo Alto Historical Association

690 E. Middlefield Road, Mountain View. By 1968 HP had $326 million in revenue and 15,840 employees. City of Mountain View.

Hewlett Packard has consistently been one of the area's largest employers and during periods of rapid expansion occupied a wide array of facilities. The company made the Fortune 500 list in 1962 with $110 million in revenue and 6250 employees.

100 Mayfield Drive, Mountain View. Mayfield Mall served as a shopping center from 1966–1984. HP converted it into an office center in the late 1980s to accommodate their growth, vacating in 2003. After sitting vacant for years, Google moved in around 2013. City of Mountain View.

Varian Associates

The Klystron. *Stanford University Archives.*

Sigurd Varian directed a demonstration of the Klystron at the ceremony to open the new Varian Building in The Stanford Research Park. Palo Alto Historical Association.

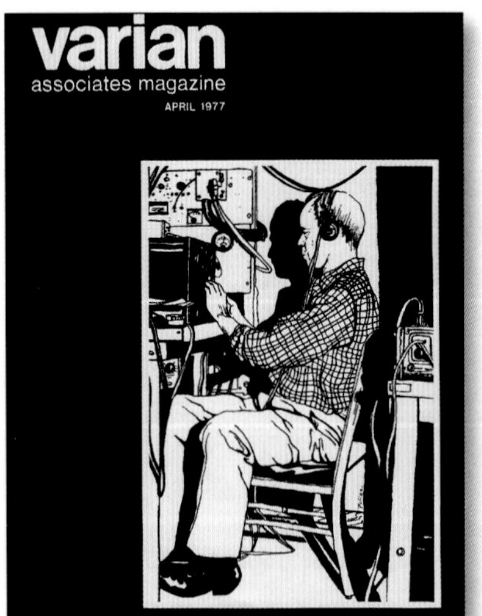

Magazine cover for Varian with a cartoon depicting the first amateur two-way 'moon-bounce' radio transmission.

Varian's initial campus, located at 3120 Hansen Way in the Stanford Research Park, is still occupied today by Varian Medical Systems. Palo Alto Historical Association.

Founded in 1948 by Stanford graduates Russell and Sigurd Varian, Edward Ginzton as well as their professor William Hansen, Varian Associates was one of the first high-tech companies in the area and maintained deep ties with Stanford. Both Fred Terman and David Packard spent time as directors. In 1953 Varian was the first tenant to occupy space in the newly formed Stanford Industrial Park. In the early years, Varian Associates subsisted almost strictly on military contracts. Their initial product was the klystron, the first tube which could generate electromagnetic waves at microwave frequencies. Their technology was instrumental in the radars used in World War II. One of Varian Associates' major contracts in the 1950s was to create a fuse for the atomic bomb. While initially supportive of military contracts, in 1958 Russell and Sigurd both expressed regret for their involvement in the development of weapons of mass destruction. The company pioneered profit-sharing, stock-ownership, insurance, and retirement plans for employees long before these benefits became mandatory. Among their early employees was bookkeeper Clara Jobs, mother of Steve Jobs. In 1999 the company was reorganized into 3 entities: Varian Medical Systems, Varian Semiconductor Equipment Associates (subsequently acquired by Applied Materials) and Varian, Inc. (subsequently acquired by Agilent Technologies).

LOCKHEED

Initially, Lockheed was located in Southern California to service the aviation industry. In 1955 they announced the creation of a missile systems division. With the goal of decentralizing their geographic locations, they purchased a 275 acre tract of land near Moffett Field in Sunnyvale. The initial construction of four buildings totaled around 250,000 square feet of office and manufacturing space. Only one year later, the new missile systems division exceeded 8,600 people in the Bay Area and so Lockheed purchased an additional 154 acres of land in Moffett Park to accommodate such rapid expansion. By 1960 the Sunnyvale population had reached 53,000—a five-fold increase over what it was just 10 years earlier. By that time, employees topped 20,000 at Lockheed Space and Missile systems. Put another way, for every family of four living in Sunnyvale 1.5 people per household worked at Lockheed. The Lockheed workforce in Sunnyvale peaked at nearly 30,000 in the mid-1980s prior to the fall of the Soviet Union. Lockheed is also credited with providing the talented workforce who eventually created some of the most profitable companies in Silicon Valley.

In 1960, rush hour in Sunnyvale was determined by Lockheed's massive employee base. Lockheed Martin. ▶

Lockheed

Aerial of Lockheed in Moffett Park in 1960. As seen here, the Lockheed campus is bordered on the east and west by orchards—on the north by the bay national refuge and on the south by the intersection of 101 and 237. Sunnyvale Historical Society.

LOCKHEED

The Polaris missile was a nuclear armed submarine-launched ballistic missile built during the Cold War by Lockheed for the United States Navy. The Polaris was first launched from Cape Canaveral in Florida in 1960. Lockheed Martin.

The RM-81 Agena was an American rocket upper stage and satellite support bus which was developed by Lockheed initially for the canceled U.S. Government's WS-117L reconnaissance satellite program. Lockheed Martin.

GTE Sylvania

The dramatic growth of the GTE Sylvania campus can be seen between the two aerials—one taken in 1953 and the other in 1963. They eventually occupied 55 acres, which was subsequently redeveloped into housing in the mid 1990s. City of Mountain View.

Built in 1963, the "Bubble" became a local landmark located on Ferguson Drive in Mountain View. GTE Sylvania used it for the fabrication, development and testing of antennas. The 60-foot high building was totally supported by air pressure from large blowers. City of Mountain View.

Sylvania was a manufacturer of electrical equipment, including radio transceivers, vacuum tubes, semiconductors, and mainframe computers. Sylvania established a campus in 1953 in Mountain View near the intersection of Whisman and Middlefield roads. At its peak, the company owned 61 acres of land in the area and employed around 2,000 men and women. In 1958 Sylvania merged with General Telephone and Electronics and changed their name to GTE Sylvania. As with many companies in the area during the sixties and seventies, their main customer was the United States Department of Defense. Today, Sylvania is probably most well known as a manufacturer of light bulbs.

Space Systems Loral

Launched in 1960, Courier was the world's first active repeater satellite. To highlight this new technology, it was used to transmit a message from President Eisenhower to the United Nations. Space Systems Loral.

Aerial of Palo Alto Campus taken in 1958. Space Systems Loral.

The company was founded as the Western Development Laboratories division of Philco Corporation in 1957 when construction began on the first building of its current campus located on Fabian Way in Palo Alto that year. The aerial above was taken in 1958 of their new headquarters located at 3825 Fabian Way, Palo Alto. After the company was acquired by Ford Motor Company in 1961 it went through a number of name changes: first to Philco-Ford; later Aeronautic Ford; later still Ford Aerospace; and finally the present name of Space Systems Loral, adopted in 1990 when it was acquired by Loral Space & Communications. It currently has around 3,000 employees in Palo Alto, a near record high. With a record of continued innovation, SS/L holds 192 active U.S. patents on satellite system technologies. Today, the company has 1.3 million square feet of office and R&D space spread over 37 buildings. The main campus sits on 77 acres.

DoD's Defense Innovation Unit At Moffett Field

Former Secretary of Defense, Ash Carter, gives a talk about the mission statement of the Defense Innovation Unit in 2016. US government via Wikimedia.

In 2015, The United States Department of Defense further solidified its relationship with Silicon Valley by locating the headquarters for the newly formed Defense Innovation Unit (DIU for short) on Moffett Field. DIU is the brainchild of former Secretary of Defense Ash Carter, a theoretical physicist who has taught at both Harvard and MIT. The mission of the DIU is to help the military by investing in emerging commercial technologies. Former Symantec CEO and Stanford alumni, Michael Brown, now runs the organization which is staffed by both civilian and military personnel. About the mission he has been quoted as saying: *"My fundamental view is that we are in a technology race. We didn't ask to be in this, but we're in it."* The Department of Defense hopes to harness some of the commercial technology coming out of Silicon Valley to solve problems currently faced by various branches of the Military. The DIU pays companies with promising technologies to engage on a pilot program basis. A few examples of current investments are highlighted on the opposite page.

Lastwall: Mountain View-based company which provides security solutions for websites and web applications. The DoD's is testing out Lastwall's authentication platform in their complex data environment.

Orbital Insight: Palo Alto-based company which provides satellite image processing at a petabyte scale (that's 1 million gigabytes worth of data) to help organizations understand global trends. The DoD is using Orbital Insight to analyze their daily imaging of the entire Earth, enabling real-time awareness of natural and man made threats.

Qool Therapeutics: Mountain View-based company which offers a non-invasive hypothermia device that helps slow inflammation and preserve cells. The Army is using Qool Therapeutics' device for combat injury mitigation.

Rhombus Power: Mountain View-based company that addresses big data problems in security, energy, and health care. The U.S. Air Force is using Rhombus Power's technology to aggregate, analyze and visualize data to assist in strategic decision-making and budget planning.

Bromium: Cupertino-based company that has developed a technology to fight against malware. The DoD is using Bromium's platform to protect its network from known and unknown malware threats.

C3 IoT: Redwood City-based company has developed a platform of software for big data, predictive analytics, AI, and IoT applications. The U.S. Air Force is working with C3 IoT to develop algorithms to help transition from standard time-based maintenance practices to more advanced predictive maintenance methods.

Airspace Systems: San Leandro-based company that offers a drone security solution capable of identifying, tracking, and autonomously removing rogue drones from the sky. The Army is working with Airspace to develop a drone which can intercept adversarial unmanned aerial systems.

The Traitorous Eight who went on to found Fairchild Semiconductor, which gave way to almost all subsequent semiconductor companies. From left: Gordon Moore, Sheldon Robers, Eugene Kleiner, Robert Noyce, Victor Grinich, Julius Blank, Jean Hoerni and Jay Last. City of Mountain View.

The same eight men years later in the same order as in the iconic photo initially taken in 1957. Semiconductor Industry Association.

Shockley Semiconductor opened its doors in Mountain View in 1956 and ushered in a new era for Santa Clara Valley. Shockley gathered eight of the brightest minds from across the country to work with him on his new silicon technology. These eight men were young visionaries and after leaving Shockley Semiconductor, they formed Fairchild Semiconductor which was not only outrageously successful on its own, it also cultivated almost every other semiconductor company in the Valley. In 1965 one of these eight men, Gordon Moore, predicted that the number of components in integrated circuits would continue to double every two years and his prediction has proven to be uncannily accurate. For example, in 1971 Intel created the revolutionary 4004 processor, the first commercially available microprocessor. It was the size of a fingernail and contained 2,300 transistors. This little chip delivered the same computing power as the first electronic computer built in 1946, which filled an entire room. Fast forward to 2010 when Intel introduced their Core Processor which held 560 million transistors, approximately 250,000 times more than the 1971 chip.

Eight: Silicon Valley

Shockley Semiconductor

William Shockley in 1956. City of Mountain View.

William Shockley was originally an employee of Bell Labs on the East Coast, a Nobel Laureate in Physics and a visiting Professor at Stanford. In 1956 he formed a company to use the element silicon for a semiconductor-based transistor instead of the then preferred element, germanium. He decided to leave the East Coast and to locate his new business on the border of Palo Alto, to be near Stanford University. Shockley pulled together a team of talented young men including Robert Noyce and Gordon Moore. This group has been referred to as the greatest collection of electronic geniuses ever assembled. While the impact of the technology was huge, the lifetime of the company was short. Shockley's managing style drove many employees away within a year. The company was never a big money maker and was sold to Clevite Transistor in April of 1960.

The Shockley diode was one of the first semiconductor devices invented. ▶

Shockley with his mentor, Lee de Forest, whose invention, the Audion, ushered in the era of vacuum tubes that preceded the Semiconductor. Semiconductor Industry Association.

Shockley Semiconductor headquarters was located at 391 San Antonio Road.

Traitorous Eight

Celebration at Rickey's in Palo Alto. Shockley, at the head of the table, celebrated the news of his 1957 Nobel Prize in Physics with his employees shortly before their infamous departure from his company. City of Mountain View.

The Traitorous Eight was a term given to the eight men who left Shockley Semiconductor in 1957, due to a conflict with William Shockley. The members of the traitorous eight were between 26 and 33 years old and included Gordon Moore, Robert Noyce, Julius Blank, Victor Grinich, Jean Hoerni, Eugene Kleiner, Jay Last, and Sheldon Roberts. With the help of Arthur Rock, the group was able to get financing from Sherman Fairchild to start their own company. The newly founded Fairchild Semiconductor soon grew into a leader of the semiconductor industry and became the incubator for almost all subsequent semiconductor companies.

The Traitorous Eight, shortly after leaving Shockley in 1957. City of Mountain View.

FAIRCHILD SEMICONDUCTOR

The Fairchild founders from left: Gordon Moore, Sheldon Roberts, Eugene Kleiner, Robert Noyce, Victor Grinich, Julius Blank, Jean Hoerni and Jay Last. City of Mountain View.

Fairchild Semiconductor was founded in 1957 at 844 Charleston Road in Palo Alto, which is now designated California Historical Landmark #1000. Fairchild Semiconductor Corporation pioneered new products and technologies which, together with an entrepreneurial spirit and manufacturing and marketing techniques, gave birth to Silicon Valley. The "Planar Manufacturing Process" invented in 1959 revolutionized the production of semiconductor devices, which enables the manufacturing of today's transistor microprocessor and memory chips. Within a few years every other transistor company copied or licensed the Fairchild Planar Process. The company saw immense growth and grew from eight employees to over 600 in less than two years. At their peak, they employed 12,000 people, made $130 million a year and held 80% of the computer market for integrated circuits. By the late 1960s the Semiconductor Division was having serious difficulty introducing new products and satisfying fast growing customer demand—due to issues with the parent company over funds needed for new facilities, and equity to retain key employees. Ultimately, there was an exodus of much of the talent who would go on to form their own ventures including: Intel, National Semiconductor, Kleiner Perkins, Sequoia Capital, Teledyne and AMD to name a few. The company has since been through a number of management teams and ownership structures but has never regained its former profitability and prominence.

A rendering of the 64,000 square foot manufacturing plant built at 545 Whisman Road in 1958. City of Mountain View.

In 1960, using the planar process invented by Jean Hoerni, Fairchild engineers managed to build a circuit with four transistors on a single wafer of silicon, thereby creating the first silicon integrated circuit.

In 1968 Fairchild built their new corporate headquarters located at 464 Ellis Street in Mountain View, known as the "Rusty Bucket." It was demolished in 1993. City of Mountain View.

The Term Silicon Valley Is Born

One of the Fairchild Business Newspapers

Electronic News

New York, N.Y., Monday, January 11, 1971

TWENTY CENTS

C-5A Subs: The Waters Get Muddier

SILICON VALLEY U.S.A.

By DON C. HOEFLER

Peripherals Firms Dig in Against IBM On Tape, Disk Price

Valentine Exiting Nat'l Semicon

Don Hoefler's newsletter, Electronic News, which popularized the term Silicon Valley. Semiconductor Industry Association.

Mountain View's iconic Wagon Wheel Tavern. Stanford University Archives.

By 1970 the now iconic Silicon Valley culture was really starting to take shape. Between 1961 and 1972, at least 60 semiconductor companies were established in Silicon Valley and most were founded by former Fairchild engineers and managers. The Wagon Wheel Tavern epitomized the open culture of the area where engineers from all different companies would come to celebrate successes, recruit staff, and exchange ideas. Many credit this open, sharing culture with aiding Silicon Valley's success. Don Hoefler was a journalist famous for his newsletter Microelectronics News, the definitive tabloid of the semiconductor industry. It was he who first published the name Silicon Valley on January 10, 1971. Hoefler was often found at the Wagon Wheel gathering information for his stories.

The Fairchildren

Shockley Semiconductor Founded in 1956 in Mountain View by William Shockley.

Founded in 1957 when the Traitorous Eight left Shockley to form Fairchild Semiconductor.

Founded in 1968 by Gordon Moore and Robert Noyce after they left Fairchild.

In 1967 National Semiconductor relocated to Santa Clara from the East Coast after hiring away five executives from Fairchild Semiconductor.

Founded in 1969 by a group of former executives from Fairchild Semiconductor.

Founded 1974 by Federico Faggin, who helped invent the 8080 microprocessor at Intel.

Founded in 1981 by Wilfred Corrigan, former CEO of Fairchild Semiconductor.

Sunnyvale Commercial Tenants In 1969

1. LOCKHEED
2. FAIRCHILD SEMICONDUCTOR®
3. National Semiconductor
4. Westinghouse
5. memorex™
6. signetics
7. AMPEX

While rather ironic today, one of the reasons Silicon Valley developed where it did was due to the abundance of affordable property in Santa Clara Valley. In the early days, chip manufacturers needed large swaths of land to build massive chip fabrication plants. Santa Clara Valley offered chip companies the perfect opportunity for expansion. Up until the 1960s, most of the Valley floor was still used as farm land. Property could be acquired and developed at a relatively low cost compared to the East Coast or even compared to land closer to San Francisco. The 1960s saw an inflection point when the population of Santa Clara County surpassed its once larger neighbor, San Francisco County. In 1960 both counties had a population of around 700,000. Today, San Francisco County has 825,000 residents whereas Santa Clara County has 1.8 million.

101
1
5
2
ARQUES AVENUE
LAWRENCE EXPY
CENTRAL EXPY
3
KIFER ROAD

NORTH SAN JOSE - 1970
TRIMBLE ROAD
SJC
101
BROKAW ROAD
880

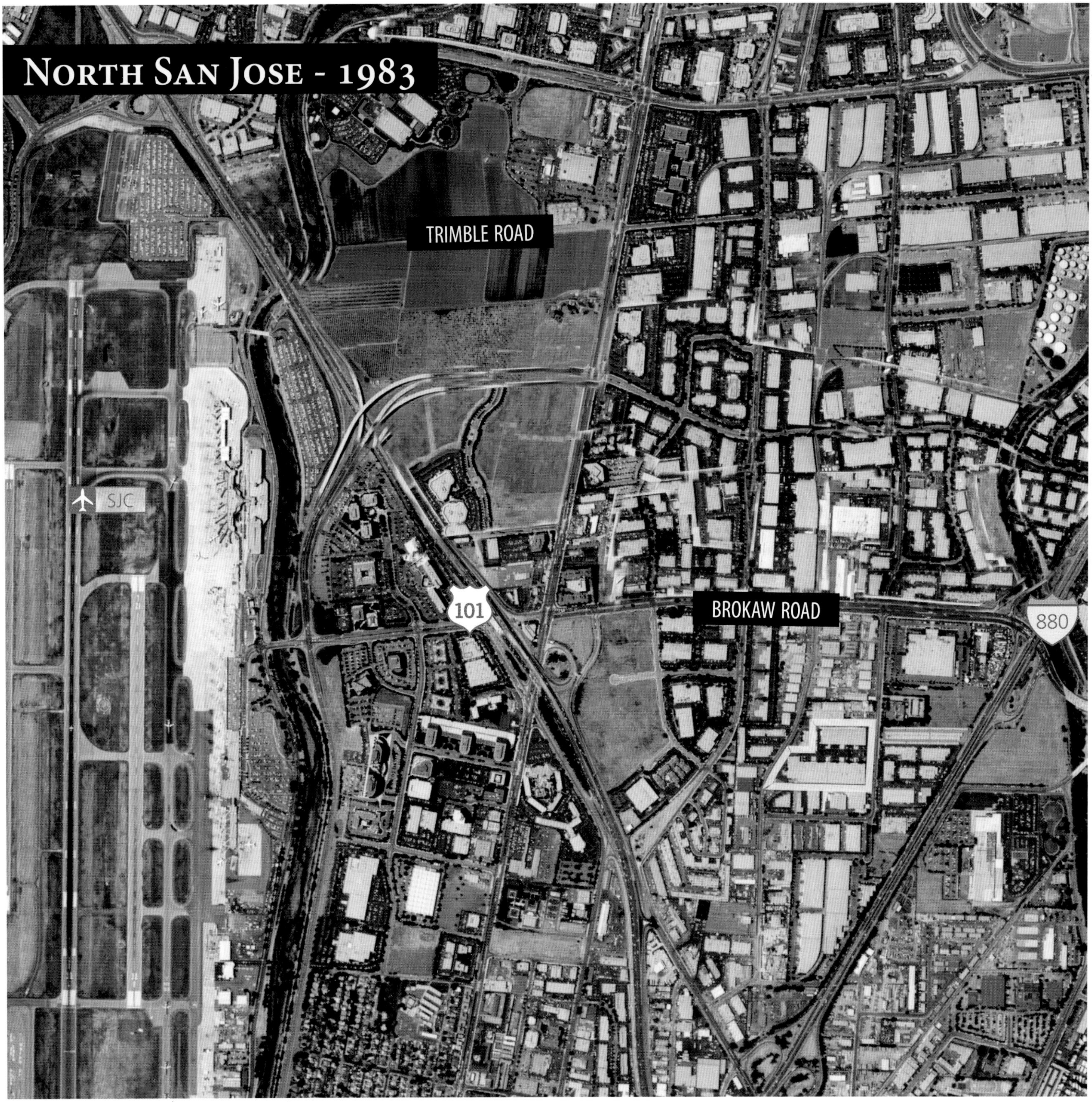
North San Jose - 1983
TRIMBLE ROAD
SJC
101
BROKAW ROAD
880

Intel

Intel employees at their new building in Santa Clara at the corner of Bowers Avenue and Central Expressway in 1971. Intel Corporation.

In 1968 Gordan Moore and Robert Noyce left Fairchild Semiconductor to found Intel using $2.5 million raised by Arthur Rock. Their company name is a combination of the words integrated and electronics. Andy Grove, who would go on to be the CEO of Intel, was their first hire. By 1971 the company had gone public raising $6.8 million and moved into a newly constructed headquarters sitting on 26 acres in Santa Clara on the corner of Central Expressway and Bowers Avenue. In 1974 Intel announced the 8080 microprocessor, perhaps the single most important product of the 20th century. At the time, it was the most complicated mass produced item ever built—each one with as many architectural features as a medium sized city and the product of programs as complicated as the Manhattan Project. It featured 4,500 transistors making it ten times faster than its predecessors. In 1978 after only ten years in business, Intel was up to 10,000 employees and by 1983 had reached $1 billion in revenue. As of 2017, they are able to pack 100 million transistors into a square millimeter of silicon chip. They remain one of the leading chip companies in the world.

Robert Noyce breaking ground for the new facility in 1970. Intel Corporation.

The Men Behind Intel

Robert Noyce is perhaps most famous for co-inventing the integrated circuit. Noyce was an east coast native and liked handicraft from early on. While in college, he attended a physics course which introduced him to the first transistors from Bell Labs. He was hooked. After college, he attended MIT to get his PhD and upon graduation went to work for Philco Ford. Two years later, he was contacted by William Shockley and moved west to join the team. After founding Fairchild Semiconductor, his first years there were the most intellectually fertile time for Noyce. Seven of his 17 patents, including his most important, the integrated circuit, date from the 18 months after the company was launched. In 1979 Noyce received the national medal of Science from president Jimmy Carter.

Gordon Moore is most famous for his predictive observations of microchip evolution, referred to as Moore's Law. He was born in Pescadero, a small beach-side community between Santa Cruz and Half Moon Bay. He graduated from Caltech then worked for Lawrence Livermore Labs before being recruited by William Shockley. Moore was one of the founders of Fairchild Semiconductor and worked there for 11 years before leaving to found Intel with Robert Noyce. In 1965 while serving as Fairchild's R&D director, he wrote an article for Electronics Magazine on the future of the semiconductor industry. He observed that over the history of computing hardware the number of transistors on integrated circuits doubled approximately every two years. This trend has continued for more than half a century and has been a driving force of technological and social change.

Andy Grove is best known for transforming Intel from a manufacturer of memory chips into one of the world's dominant producers of microprocessors. During his tenure as CEO, Grove oversaw a 4,500% increase in Intel's market capitalization from $4 billion to $197 billion, making it the world's 7th largest company, with 64,000 employees. Grove was Gordon Moore and Robert Noyce's first hire. He served as Intel's president starting in 1979, its CEO in 1987, and its Chairman and CEO in 1997. He relinquished his CEO title in May 1998, having been diagnosed with prostate cancer although remained chairman of the board until November 2004. Under his leadership, most of the company's revenues were reinvested in research and development, along with building new facilities, in order to produce improved and faster microprocessors.

Robert Noyce and Gordon Moore in 1984. Intel Corporation.

From left to Right: Andy Grove, Robert Noyce and Gordon Moore. Intel Corporation.

National Semiconductor

Intel founder Robert Noyce, and the CEO of National Semiconductor, Charlie Sporck, playing chess. Many attribute the success of the Valley on the information sharing and problem solving that takes place even between competing companies. Texas Instruments.

Charlie Sporck, CEO of National Semiconductor from 1967 to 1991, was also known for his love of cigars. Texas Instruments.

National Semiconductor was founded in Connecticut in 1959. In 1967 it shifted its headquarters to Santa Clara to attract local talent employed at companies like Fairchild and HP. One of the Fairchilders they recruited was Charles Sporck, the operations manager at Fairchild directly under Bob Noyce. Charles Sporck was appointed President and CEO of National. Sporck steered the company towards the growing market for commercial and industrial applications, thereby lessening their dependence on the highly competitive military and aerospace market. Additionally, Sporck's cost control efforts led him to be one of the pioneers in the semiconductor industry to outsource final manufacturing of integrated circuits, especially to Southeast Asia. Under his management, National Semiconductor became the first semiconductor company to reach $1 billion annual sales in 1981. Texas Instruments acquired National Semiconductor in 2011.

Advanced Micro Devices

Jerry Sanders founded AMD in 1969 and served as chairman and CEO until 2002. In 1961 Jerry moved West from his home in Chicago when he was recruited to Fairchild Semiconductor. He quickly rose to a succession of management positions and seemed headed for the top. But new management came in at Fairchild and Sanders' style made them uneasy. They fired him. Sanders got together with other ex-Fairchilders to start AMD. Sanders wrote the business plan and developed a program to go out and raise money. Sanders jokes that while Bob Noyce often told the story that it took him five minutes to raise five million dollars it was the other way around for Sanders—taking him five million minutes to raise five dollars. But Sanders was determined—he raised $1.5 million and in 1969 AMD was born. Sanders is regarded as one of the most colorful, clever and bold in the generation that founded the semiconductor industry and Silicon Valley. Jerry was an engaging leader with the motto, "Put people first, products and profits will follow." AMD was the first company to give every employee shares of its stock and the first company to participate in profit sharing. He had a taste for outlandish and theatrical spirit-rousing. He would materialize from a cloud of smoke at a sales meeting wearing a pink jumpsuit and stare down from in-house movie posters posing as "Illinois Jerry," a character fighting the evil forces against AMD.

Jerry Sanders in 1980. AMD.

The AMD management team at the ground breaking ceremony for their new Sunnyvale headquarters on Thompson Way in Sunnyvale in 1969. AMD.

The completed 901 Thompson building at the right along with the later two story expansion at 902 Thompson, circa 1970. AMD

LSI Logic

LSI founders from left to right: Mick Bohn, Wilf Corrigan, Rob Walker, and Bill O'Meara.

LSI Logic was founded in Milpitas in 1981 by Wilf Corrigan as a semiconductor company after he left his post as CEO of Fairchild Semiconductor in 1979. The other three founders were Bill O'Meara, Rob Walker and Mick Bohn. Corrigan was CEO for LSI for over 20 years until 2005. The initial $6 million round of funding was led by ex-Fairchilder Don Valentine who by that point had started Sequoia Capital. The firm went public in 1983—netting $153 million, the largest tech IPO up to that point. LSI developed the industry's first line of ASIC (application-specific integrated circuits) which let customers create custom chips by use of leading-edge proprietary CAD tools. Renamed LSI Corporation, the company today focuses on chips that handle the massive amounts of data generated by our digital society.

Applied Materials

The first office of Applied Materials in 1967—a 750 square foot industrial unit in Mountain View. Applied Materials.

Applied Materials founder, Michael McNeilly.
Applied Materials

A rendering of the new Applied Materials campus on Bowers Avenue in 1974. Applied Materials

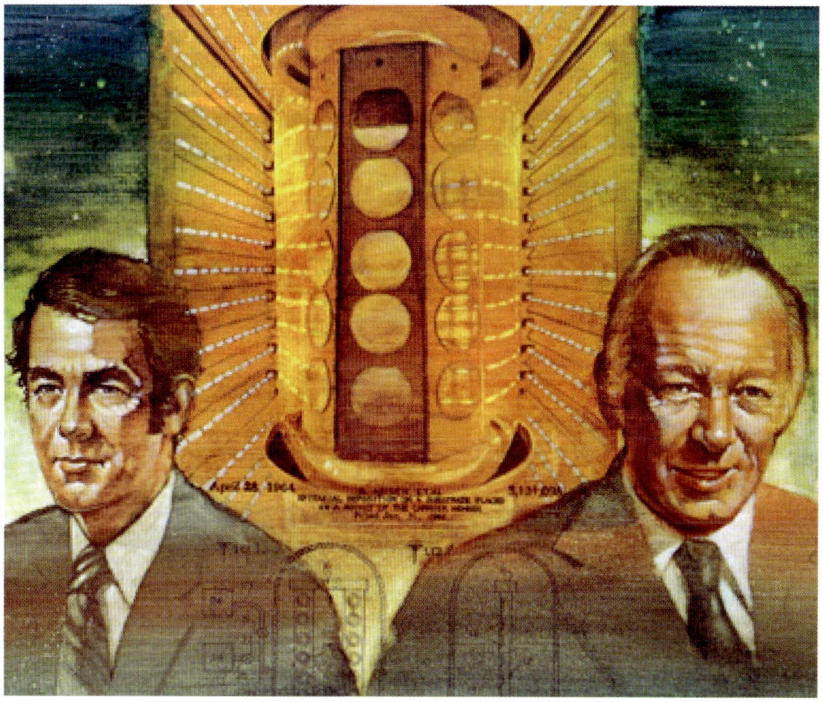

Michael McNeilly & Walter Benzing pioneered epitaxial deposition equipment at Applied Materials. SEMI.

In 1972 the AMS 2000/2001 Continuous Silo Reactor system changed the concept of equipment architecture from "batch-at-a-time" to high throughput continuous processing. Applied Materials.

Applied Materials, originally known as Applied Materials Technology, was founded in 1967 by Michael McNeilly at the ripe old age of 28 with a $7,500 loan from his father-in-law in a 750 square foot industrial unit in Mountain View. The initial goal of Applied Materials was to supply the rapidly growing semiconductor companies with the materials and equipment needed for their manufacturing processes. While most of the large semiconductor companies had their own internal process development capabilities, the semiconductor industry was spawning many new device start-ups who needed equipment for their ventures. McNeilly credits the early success of the company to the relationship he had with the management team of Fairchild through informal meetings at the Wagon Wheel and playing basketball for the Fairchild team. The early board of directors and investors in Applied Materials included Bob Noyce, Tom Perkins, Jean Hoerni and Charlie Sporck—all early members of Fairchild Semiconductor. Once Applied raised the capital they needed to grow the business, they moved into an 18,000 square foot building on San Ysidro Way in Santa Clara, directly across the street from National Semiconductor, and grew the business to 100 employees and revenues of about $10 million in their first year there. McNeilly instituted a policy early on that any prototype system built for a customer must also be installed in a lab in the Applied Materials space, which was the first cleanroom demo lab in the industry. In 1972 Applied Materials dropped 'Technology' from their name and went public. In 1974 they relocated from San Ysidro to a larger building on Bowers Avenue. Today, Applied materials has nearly $9 billion in annual sales and holds over 10,000 patents.

Cypress Semiconductor

Rodgers earned the moniker as the "Bad Boy of Silicon Valley" for calling out his competitors in the semiconductor industry for being dinosaurs—he accused them of calling for political protection instead of innovating and investing in new technology. *Used with permission of Bloomberg L.P.*

Cypress founder and Wisconsin native, T. J. Rodgers is a Green Bay Packers fan and owner of Clos de la Tech winery which he founded with his wife, Valeta Massey. T.J. Rodgers

Cypress Semiconductor CEO as of 2016, Hassane El-Khoury.
Cypress Semiconductor

T.J. Rodgers co-founded Cypress Semiconductor in 1982. He served as CEO until his retirement in 2016. In the early 1980s, there were two primary types of memory chips for semiconductor products: bipolar chips (fast but expensive) and NMOS chips (slow but less costly). Rodgers founded Cypress Semiconductor to design, develop and sell a new type of chip which would offer the high performance of bipolar but at a price comparable to NMOS. Cypress Semiconductor released their first product in 1984 and went public 2 years later. Around this time, Silicon Valley as a whole was facing a serious threat to its very existence as the quality of the semiconductor chips from Japan far exceeded those made in the Valley. Speaking in front of the congressional committee tasked at managing this threat, Rodgers assured the room that the Valley would ultimately prevail based on innovation. Years later, Cypress Semiconductor purchased the once famed Japanese chip maker Fujitsu Semiconductor (through its acquisition of Spansion). Rodgers was also known for over-the-top publicity stunts. Towards the end of the cold war, he published a full-page advertisement inviting Soviet Premier Mikhail Gorbachev to come to Cypress Semiconductor's headquarters in San Jose to see "capitalism in motion." The secret service nixed Gorbachev's visit but 10 senior Soviet ministers did make the trip. Under Rodgers' 34-year role as CEO, Cypress Semiconductor acquired 29 companies and received over 5000 patents. In 2016, Rodgers selected 36-year-old Hassane El-Khoury to succeed him as CEO. He was relatively young and with a background in the automotive industry as opposed to semiconductor chips. This non-traditional choice turned out to be an excellent bet. El-Khoury has guided the company to focus on markets growing faster than the overall semiconductor industry, including chips for Internet of Things (IoT) devices as well as micro-controllers used in the automotive industry. Cypress achieved record revenue of $2.33 billion in 2017—a 20 percent year-over-year increase. Today, Cypress Semiconductor has around 7,000 employees with a market cap over $6 Billion.

Synopsys

Synopsys (SYNthesis and OPtimization SYStems) was founded in 1986 by Dr. Aart de Geus along with engineers Bill Krieger and David Gregory. The founders wanted to create software that would accelerate the process of designing chips. Prior to electronic design automation (EDA), all chips were designed manually, similar to an architect drawing blueprints by hand. Synopsys' original flagship product was Design Compiler, a set of software tools that automated the design of integrated circuits. Synopsys is now the world's leading provider of EDA technologies for designing and verifying advanced chips, and for designing the next-generation processes and models required to manufacture those chips. Almost any high-tech device seen today—from smartphones to self-driving cars—contains chips designed using Synopsys software.

Synopsys co-founders William Krieger, Aart de Geus and David Gregory. Synopsys

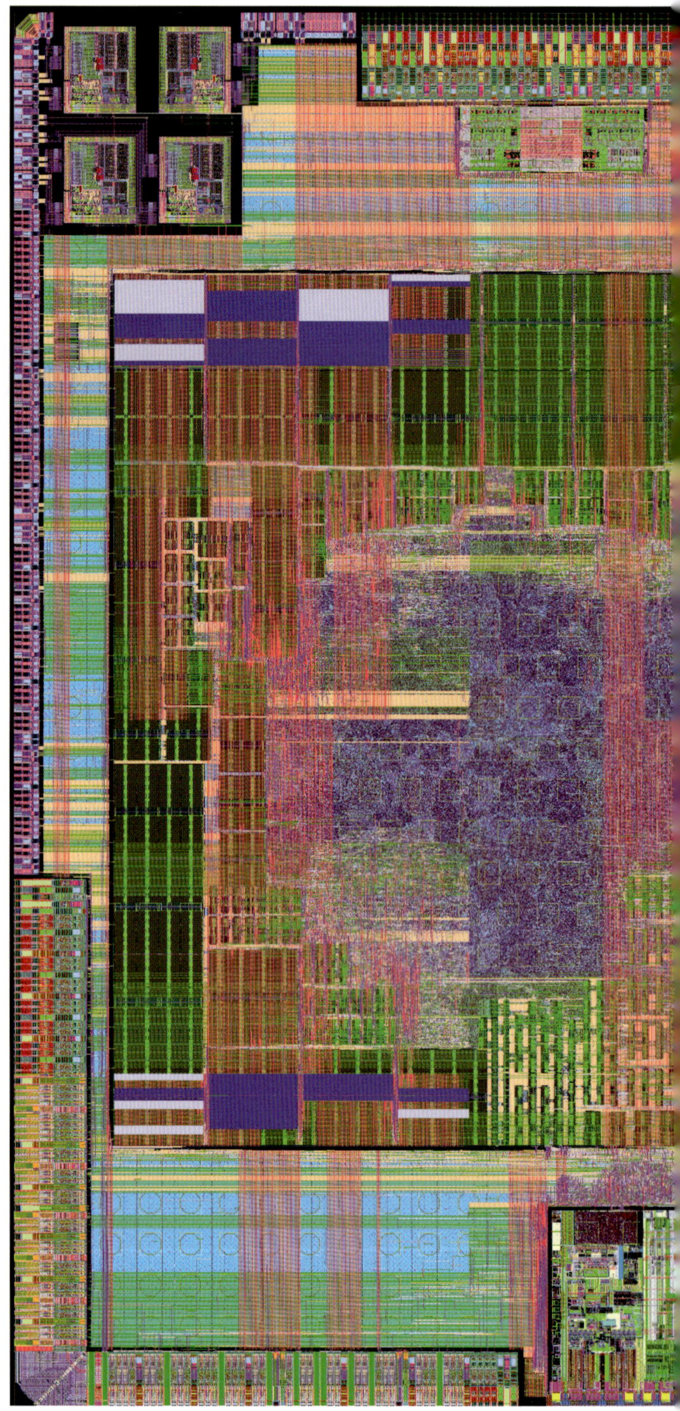

Synopsys' DesignWare® ARC® Processor IP is used by compani digital home, mobile, automotive, industrial and Internet of Thin

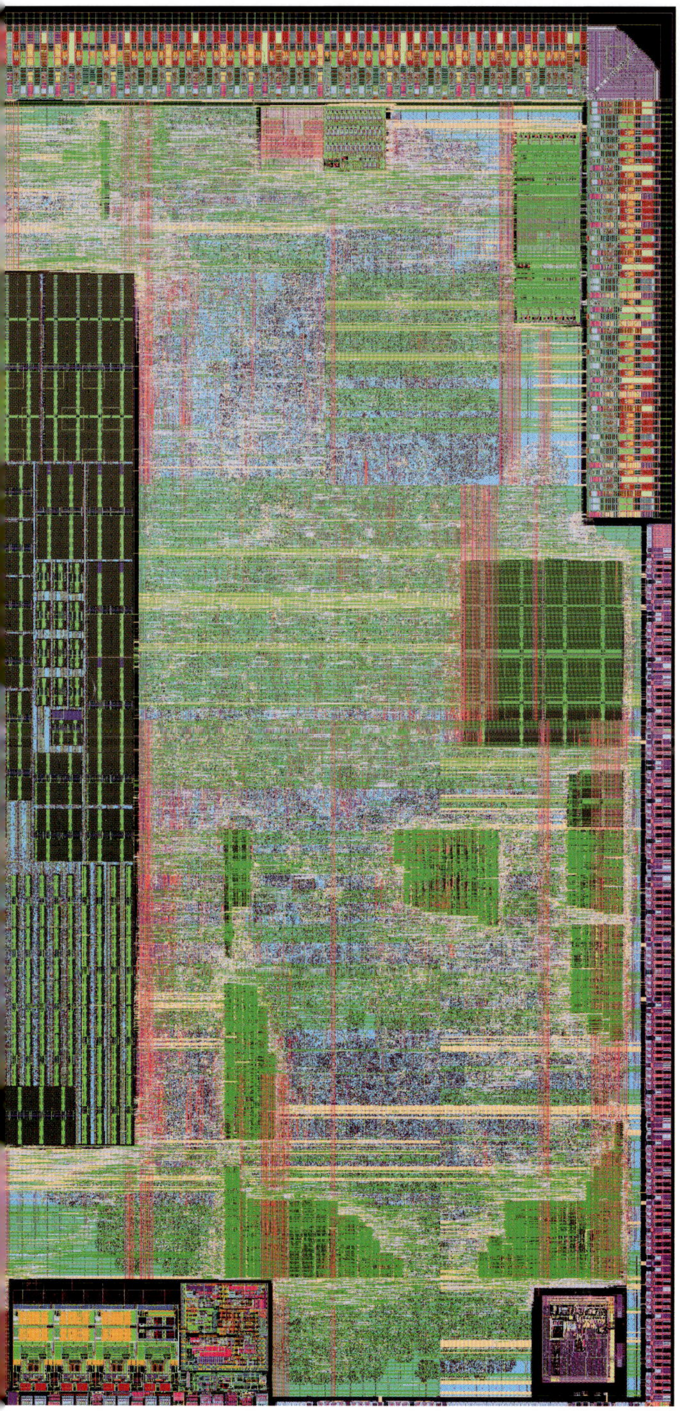

orldwide in chips for applications varying from storage,
oT). *Synopsys*

Synopsys co-CEOs Aart de Geus and Chi-Foon Chan. Synopsys

In addition to advancing the field of EDA, Synopsys leads the market in silicon IP—reusable blocks of logic, interface, and processor design. Synopsys ARC processors are used in billions of chips for applications including storage, digital home, mobile, automotive, and IoT. Synopsys has recently emerged as the leader in application security testing. Shortly before the company went public in 1992, Dr. Chi-Foon Chan joined the team. He was appointed president in 1998 and in 2012 he joined Aart as a co-CEO. The pair continues to share the role as co-CEO. The company is headquartered in Mountain View, with more than 13,000 employees worldwide and $3 billion in annual revenue. Synopsys technology is bringing digital intelligence to customers around the world, helping them design the future of Smart, Secure Everything.

NVIDIA

NVIDIA's early team in front of their first office located at 1226 Tiros Way in Sunnyvale. NVIDIA

NVIDIA was founded in 1993 by Jensen Huang, Chris Malachowsky and Curtis Priem. The trio believed that graphics-based computing would be a way to meet the processing power requirements of the exploding video game market. As opposed to general purpose processors from Intel and others which are built to handle a wide range of unpredictable and complex tasks, NVIDIA's chips execute thousands of relatively simple tasks simultaneously. The company's unusual name took its inspiration from the founders' convention of adding "NV"—short for "next version"—to all their files, as well as the word "invidia," which is Latin for envy. In 1999, NVIDIA invented and coined the term Graphics Processing Unit (GPU for short) which is a single chip that can process large sets of data. NVIDIA went public later that year. According to an Investopedia article published in October 2018, "If you bought 100 shares in 1999 after four stock splits you would now be the owner of 1,200 NVDA shares, and your initial capital with reinvested dividends would have grown to $274,231."

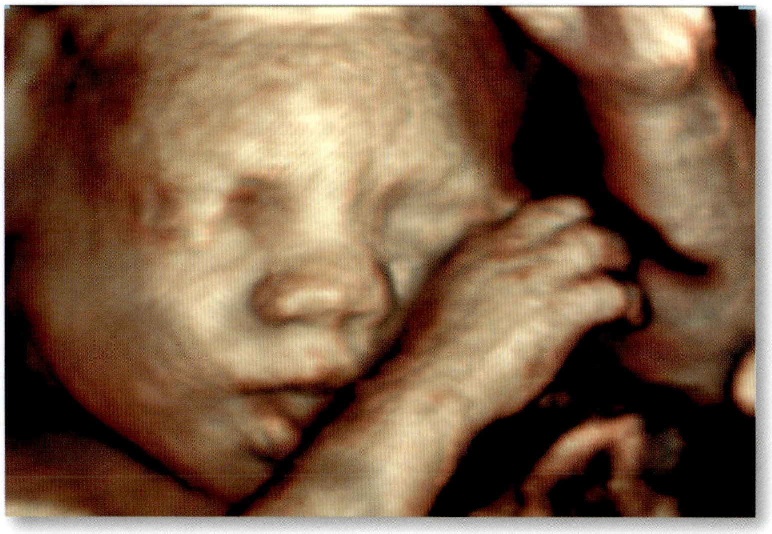

In 2009 NVIDIA partnered with Siemens to create the world's first 3D ultrasound. GPUs play an important role in medical imaging by processing large amounts of visual data. NVIDIA

NVIDIA CEO Jensen Huang in 2018 with a self-driving car demo, during which an NVIDIA employee was "teleported" into a virtual car to remotely drive an actual car. NVIDIA

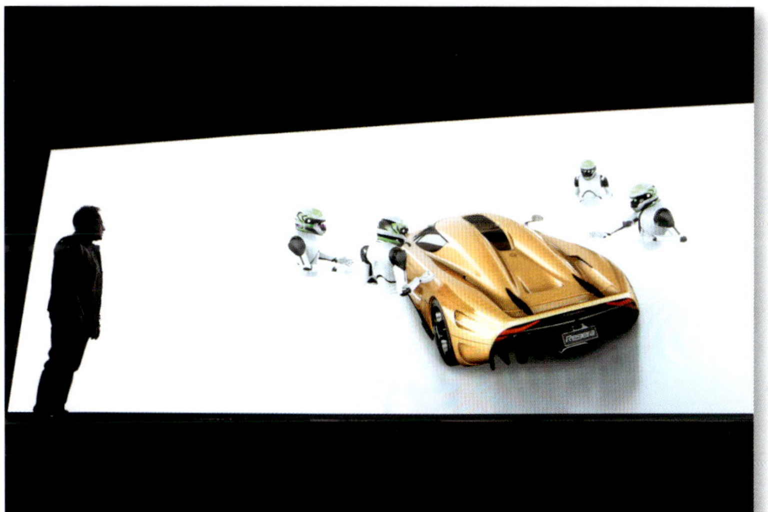

In 2017 CEO Jensen Huang introduced NVIDIA Holodeck, a platform that allows content creators to collaborate and interact with their designs in photo-realistic detail in a simulated VR environment. NVIDIA

NVIDIA's 500,000 sq. ft. headquarters in Santa Clara. The building was designed featuring triangles, the fundamental building block of computer graphics. The triangle is the most stable of shapes because it doesn't easily deform and can balance the stretching and compressive forces inside a structure. NVIDIA

In 2007, Forbes magazine named NVIDIA Company of the Year. The year 2012 was the big bang of AI, or Artificial Intelligence. CEO Jensen Huang recognized the potential of GPUs in deep learning and the company modified its technology to assist developers and AI researchers with their work. NVIDIA has established market-leading positions beyond its roots in gaming, extending into self-driving cars, robotics, medical imaging and virtual reality.

Sand Hill Rd

By the 1950s Santa Clara Valley was already becoming defined by its culture of entrepreneurship. However, in the 1950s and into the early 1960s, there was no such thing as venture capital, which meant a young entrepreneur who wanted to start a company essentially had two options for financing. The first option was to secure a decent sized military contract, which paid for both the upfront R&D costs and the purchase of the final product. Hewlett Packard and Varian followed this approach: they secured military contracts and depended on them for their survival. The second option was to woo an established company to sponsor the new venture. Shockley Semiconductor and Fairchild Semiconductor were funded in this manner as each was initially backed by established corporations (Beckman Instruments and Fairchild Camera, respectively). However, in the late 1950s things started to change as local companies began to go public: Varian in 1956, Hewlett Packard in 1957, and Ampex in 1958. To potential investors, this offered a means of liquidity. Also in 1958, the SBIC Act was passed which provided 4:1 leverage for any investment institution that provided funds to small innovative companies. This meant that for every $1 private investors lent to a small business, the U.S. Government would invest an additional $4. A large concentration of SBIC investment groups set up in the Bay Area to tap into the emerging microwave and semiconductor start-up companies. The exponential growth of the technology sector in Santa Clara Valley helped propel Sandhill Road in Menlo Park into the premier venture capital region of the world. This chapter explores some of the pioneers.

Nine: Venture Capital & Sand Hill Road

Sand Hill Road

Sand Hill Road in 1904, facing west. Ford Land Company.

Sand Hill Road in 1970, taken from the air over highway 280 and focused on the newly constructed 3000 Sand Hill Circle development in Menlo Park. Ford Land Company.

Arthur Rock

- Scientific Data Systems
- Fairchild
- Intel
- Teledyne
- Apple

Arthur Rock was born in 1926 in New York and attended Syracuse University for his undergraduate degree and later Harvard Business School. Upon graduation from Harvard, Rock joined the Wall Street investment firm, Hayden, Stone & Co. In 1956 when the "Traitorous Eight" were looking to leave Shockley Semiconductor, it turned out that one of them, Gene Kleiner, had a father with an account at Hayden, Stone & Co. Gene wrote a letter to his father's investment house saying, "A lot of us like working together. Do you think there's a company that would hire the whole group?" It was Arthur Rock who was instrumental in getting "The Eight" not to join another company but to start their own. He approached all of the wealthy clients he had in the attempt to raise funding for their new venture, but almost everyone said no. When close to giving up, he found an investor: Sherman Fairchild, who decided to invest $1.5 million in the new start-up as a way to diversify the product line for Fairchild Camera. After the wild success of Fairchild Semiconductor, Arthur Rock decided to chase the wave and moved to California in 1961. He teamed up with Thomas Davis Jr. and founded California's first true venture capital firm: Davis & Rock. The investment company operated until 1968 and during those seven years they funded 15 new companies. The firm hit it really big with an investment in Scientific Data Systems, or SDS — one of the first companies to use microchips in computers. Davis and Rock invested close to $260,000 and in only eight years Xerox purchased SDS for almost $1 billion. In 1968 when Robert Noyce and Gordon Moore split from Fairchild to form Intel, Rock raised the $2.5 million needed to fund the creation of Intel. In fact, he invested $300,000 of his own money and acted as the first president of the company. Another major milestone for Rock was meeting Steve Jobs and Steve Wozniak and his subsequent investment in Apple in 1978. At the time, he bought 640,000 shares of Apple at nine cents per share for a total investment of $57,000. Three years later Apple went public and Rock's $57,000 became $14 million.

Tommy Davis and the Mayfield Fund

- Silicon Graphics
- Compaq
- LSI Logic
- 3COM
- SanDisk

Tommy Davis Jr. was born in 1912 in Cincinnati and earned his Bachelor of Arts and Law degrees from Harvard University. After serving in World War II, he began working for the Kern County Land Company in California's Central Valley. With the technology boom that was beginning to take hold in Santa Clara Valley, he decided to get out of land development and use his own money to make angel investments in fledgling technology companies. In 1963 he teamed up with Arthur Rock and after some wildly successful investments together, they decided to go their separate ways. In 1969 Tommy Davis teamed up with Wally Davis (no relation) and with an investment from Stanford University, they founded the Mayfield Fund. The first fund was around $3 million. Since then Mayfield has raised over $2.8 billion of investor commitments across 13 private equity funds. They have invested in more than 530 companies resulting in over 100 Initial Public Offerings and more than 100 mergers and acquisitions. The firm is still headquartered on Sand Hill Road in Menlo Park.

The Draper Dynasty

William Draper Jr. in 1978.
Draper Fisher Jurvetson.

From left to right: Bill Draper III, Tim Draper and William Draper Jr. Taken in 1978.
Draper Fisher Jurvetson.

William Draper III in 1959.
Draper Fisher Jurvetson.

Steve Jurvetson, Tim Draper and John Fisher in 1985 upon the founding of Draper Fisher Jurvetson. Draper Fisher Jurvetson.

- LSI Logic
- Apollo Computer
- Quantum
- Xidex
- Measurex

- Hotmail
- PLX Technologies
- Skype
- Baidu
- Digidesign

The Draper family dynasty began with William Draper Jr. in the late 1950s. William Draper Jr. grew up on the East Coast and had a background in investment banking. In 1958 Draper got together with Rowan Gaither, the founder of the Rand Corporation, and Frederick L. Anderson, a retired Air Force general. Together they formed what was the first venture capital firm to be structured as a limited partnership. Draper, Gaither & Anderson remained in existence through 1966.

William's son, Bill Draper III, followed in his father's footsteps. He graduated from Yale University in 1950, a year after George H. W. Bush. Bill Draper went on to attend Harvard Business School and studied under professor Georges Doriot, who is credited with mentoring many of the early venture capitalists. In 1959 Bill Draper came to work as an associate at his father's newly formed firm. He canvased the area, approaching any storefront with the word technology in its name to ask if they wanted some capital in exchange for equity. A few years later in 1962, Bill left his fathers firm to co-found the venture capital firm Draper & Johnson Investment Company with his good friend Pitch Johnson. In 1965 Draper founded Sutter Hill Ventures, which fifty years later, remains one of the top venture capital firms in the country. During his twenty years as the senior partner of Sutter Hill, Draper helped to organize and finance several hundred high technology manufacturing companies including Apollo Computer, Dionex, Integrated Genetics, Quantum, Xidex, Measurex, Hybritech, and LSI Logic. In 2005 he received the Vision Award from SD Forum and was inducted into the Dow Jones Venture Capital Hall of Fame. In 2006 he received the Silicon Valley Fast 50 Lifetime Achievement Award.

Bill's son, Tim, continued in the family legacy and in 1985 started his own venture capital firm, Draper Fisher Jurvetson. He has a BS in Electrical Engineering from Stanford University and an MBA from Harvard Business School. He has been named #52 of the 100 most influential Harvard alumni, and #7 on the Forbes Midas List. He was named the #1 venture capital deal maker in 2008 by Always-On. His investments include Skype, Baidu, Parametric Technology, Hotmail, PLX Technologies, Digidesign and hundreds of others.

Gene Kleiner & Tom Perkins

From left to right: Brook Byers, Frank Caufield, Tom Perkins, and Eugene Kleiner. KPCB.

- **Tandem Computers**
- **Genentech**
- **Applied Materials**

In 1972 Gene Kleiner, an original founder of Fairchild Semiconductor and Thomas Perkins, a leader of Hewlett Packard's early computer hardware division, formed their own venture capital firm using $8 million from a private investor. In the beginning many of their deals lost money and after a couple of years it was apparent they needed to do something different. That something happened to come in the form of an opportunity to invest in Tandem Computers under the leadership of larger-than-life Texan, Jimmy Treybig. It was an interesting time in the computer market in 1974: IBM controlled the mainframe market and there was little room for competitors. But Tom Perkins saw a need in the market for stable mainframe computers that would never fail, regardless of the environment. His intuition paid off. The development of Tandem's no fail mainframe led to its exponential growth. Tandem made the Fortune 500 list by 1982. Perkins' original investment of $1.4 million in 1974 turned Tandem into a $3 billion company by the time it sold to Compaq in 1997.

Don Valentine & Sequoia Capital

- Apple Computer
- Atari
- LSI Logic
- Cisco Systems

Don Valentine grew up in a suburb of New York City and got his start in technology working in vacuum tube sales for Sylvania. In 1959 Valentine realized the future of the technology market was in semiconductors, and that meant California. He left the East Coast to begin work for Fairchild Semiconductor. He worked his way through the ranks of the sales team, where he is credited with hiring many of the bright young men who went on to become industry leaders—including Jerry Sanders (founder of AMD), Jack Gifford (founder and CEO of Maxim) and Mike Markkula (initial investor and CEO of Apple). In 1968 Don Valentine joined Charles Sporck as a defector from Fairchild to help found National Semiconductor. In 1972 he left the semiconductor business and founded Sequoia Capital with an $8 million investment. One of his first investments was in Atari in 1974. A few years later, it was Valentine who put Steve Jobs and Steve Wozniak in touch with their initial investor and CEO, Mike Markkula. Since then he and his Sequoia Capital Partners have participated in the financing of over 500 technology companies. Some of the more famous investments include: Atari in 1974; LSI Logic in 1981; Electronic Arts, which was started in Sequoia's office in 1982; and Cisco Systems in 1987.

Venture Funding By Region 2002 - 2018

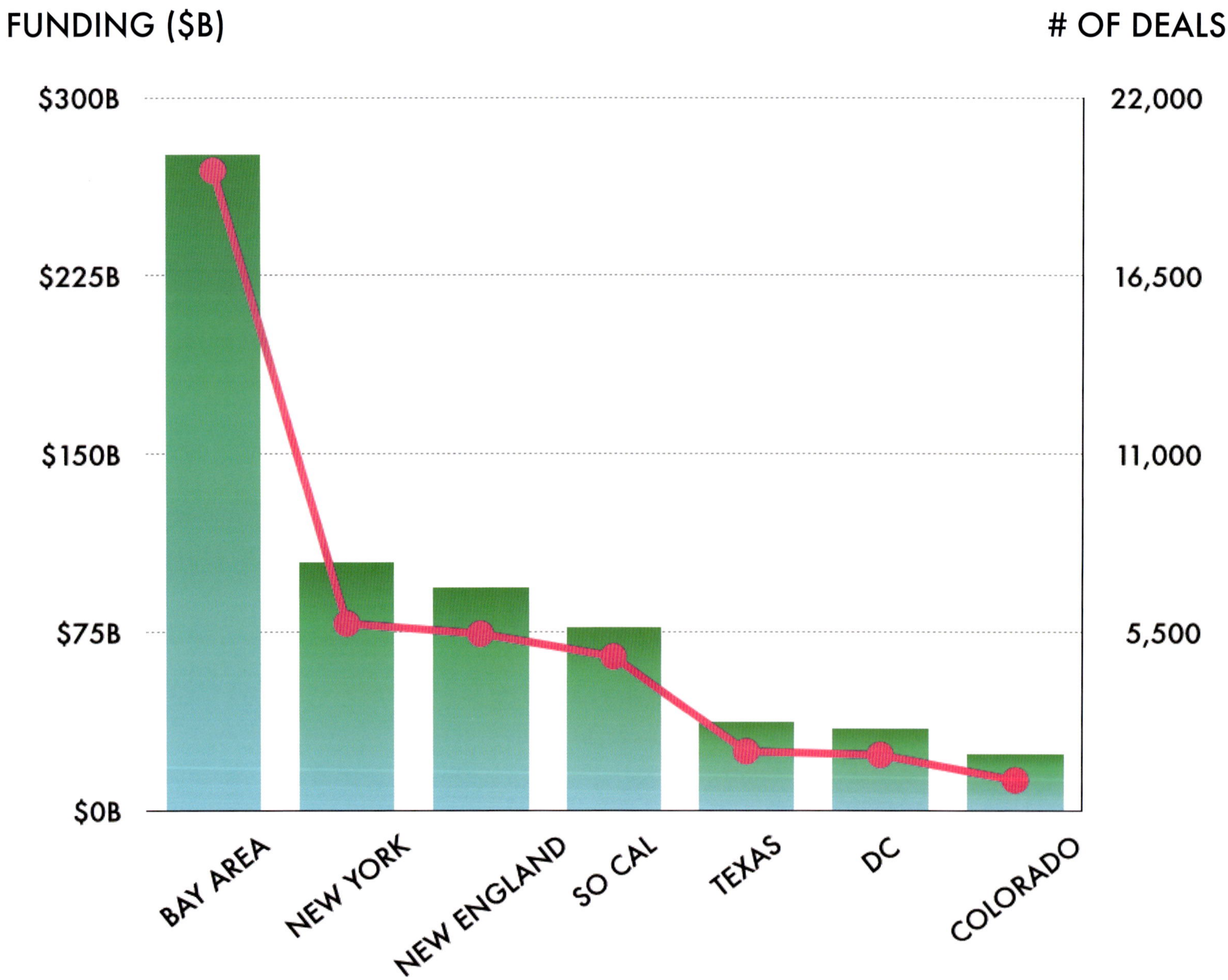

The chart above reflects data complied by Price Waterhouse Coopers on the venture capital industry dating back to 2002. The Bay Area (including San Francisco & Silicon Valley) has seen over $270 billion in investments placed into approximately 20,000 start-up companies over the last sixteen years. That is nearly equivilent to the amount of money invested in the next six regions combined.

Buck's of Woodside

Interior of Buck's of Woodside. Jamis MacNiven

Opened in 1991, Buck's is known for its wild decor as well as being the ultimate power lunch spot. Decorated with model planes hanging from the ceiling and walls filled with eccentric decor, it is the restaurant's proximity to Sand Hill Road that has made it the go-to spot for entrepreneurs and investors looking to strike a deal. A few notable encounters incude: Marc Andreessen used Buck's as the spot to pitch Netscape to venture capitalist John Doerr; a shared meal between Sabeer Bhatia & venture capitalist Steve Jurvetson launched Hotmail; Elon Musk and Peter Thiel held early PayPal meetings at a back table; Jerry Yang met with Bill & Tim Draper here while trying to raise capital for Yahoo!; and the founders of Tesla parked a bright yellow prototype out front in an effort to raise capital by offering offering chunks of equity in the company for $50,000. More tales from the colorful restaurant owner Jamis MacNiven can be found in his book: *Breakfast at Buck's: Tales from the Pancake Guy.*

The program development room for the SAGE system computer in the 1960s. IBM.

Network Room for the SAGE Computer System. IBM

In the early days of computers, due to their size and cost, only very large corporations and the U.S. Government could afford such machines. By far, the Defense Department was the computer's largest consumer. In the 1960s it is estimated that the Department of Defense owned more than 75% of all the computers in existence. A great example of a computer of the 1960s was the SAGE (Semi-Automatic Ground Environment), a multi-billion dollar state-of-the-art computerized air defense system acquired by the Air Force, in conjunction with MIT and IBM, at the beginning of The Cold War. Starting in 1958 the Air Force built around 20 SAGE centers across the U.S. Each center consisted of a four-story windowless building used to house all of the equipment related to the computer. The system cost the equivalent of $1.89 billion in today's dollars, occupied 113,000 square feet of floor space and contained 50,000 vacuum tubes.

Ten: Mainframe Computers

IBM

IBM's General Products Division in San Jose. Voted "One of America's ten most beautiful plants" in the 1960s. IBM.

IBM's Building 25. Built in 1957 in the Edenvale district of San Jose, it was designed to be modernistic and was state of the art at the time. IBM.

An aerial view of IBM's South San Jose campus in 1974. IBM.

BUSINESS INTERNATIONAL MACHINES

In 1952 IBM opened a West Coast laboratory in San Jose. The purpose of the new facility, deliberately removed from the mainstream IBM labs, was to seek and develop radically new information processing capabilities. When the operation grew from eight employees to 60 within the first year, IBM purchased 190 acres of land in the Edenvale area of South San Jose. Called building 25, the Cottle Road facility was designed to be modern and cutting edge. One of the most significant research projects in the early days was the invention of the hard disk drive. In 1960 an IBM 305 RAMAC was the first computer to tabulate scores at the Olympic Games. By the end of the 1960s IBM had close to 80% of the computer market and was one of the most successful and admired corporations in the world.

IBM

Introduced in 1956, the IBM 305 RAMAC (Random Access Memory Accounting System) was an electronic general purpose data processing machine that maintained business records on a real-time basis. The 305 RAMAC was one of the last vacuum tube systems designed by IBM, and more than 1,000 of them were built before production ended in 1961. IBM.

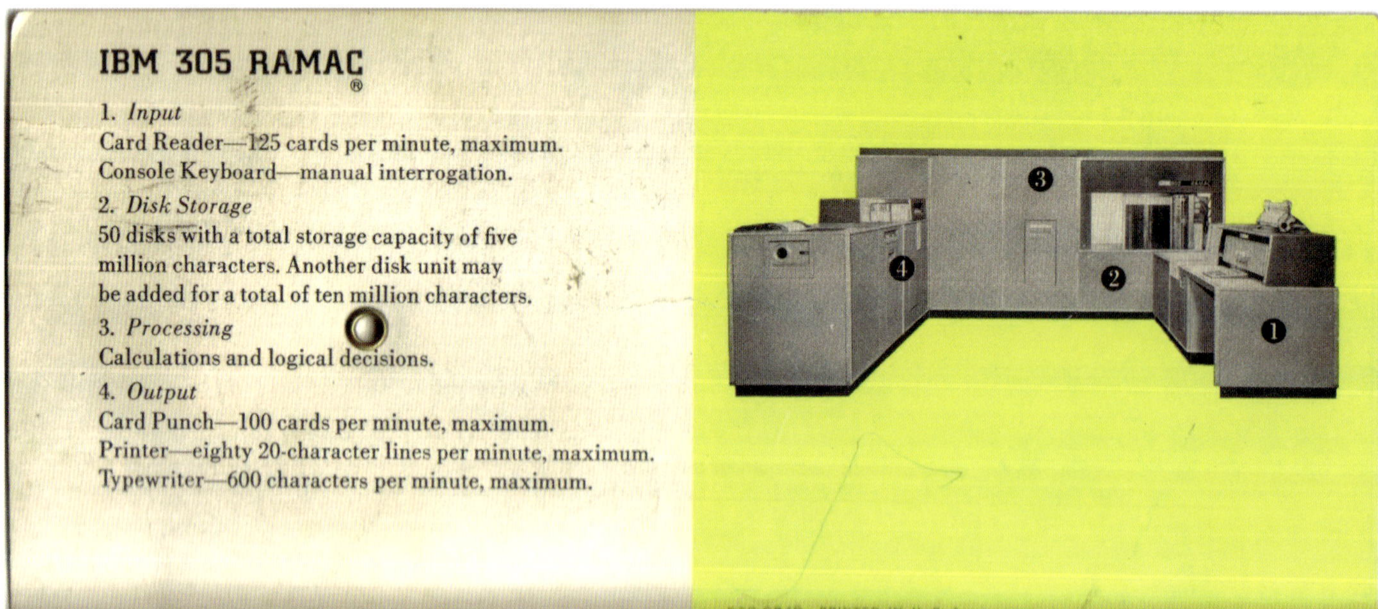

IBM 305 RAMAC®

1. *Input*
Card Reader—125 cards per minute, maximum.
Console Keyboard—manual interrogation.
2. *Disk Storage*
50 disks with a total storage capacity of five million characters. Another disk unit may be added for a total of ten million characters.
3. *Processing*
Calculations and logical decisions.
4. *Output*
Card Punch—100 cards per minute, maximum.
Printer—eighty 20-character lines per minute, maximum.
Typewriter—600 characters per minute, maximum.

520-0840 PRINTED IN U. S. A.

In 1960 IBM's 305 RAMAC scored the Winter Olympic Games which were held in Lake Tahoe. The RAMAC also tallied votes for the presidential election that same year. IBM.

Tandem Computers

Tandem Founders Mike Green, Jack Loustaunou, Jim Katzman, and Jimmy Treybig. Gaye Clemson.

In 1974 an article came out that claimed there would never be another computer company because IBM had eaten up the entire mainframe market. Incidentally, it was only months after this article was published that Tandem Computers was founded and soon became the dominant manufacturer of fault-tolerant computer systems for ATM networks, banks, and stock exchanges. Tandem Computers was founded by Jimmy Treybig, an eccentric Texan who first saw the market for ultra-reliable systems that would provide maximum uptime and zero data loss. Working in the commercial computing division at Hewlett Packard at the time, he mentioned this to his employer but they were not interested in developing that niche. Initial venture capital investment came from the investment firm, Kleiner Perkins. Treybig worked for Tom Perkins during Perkins' tenure at HP so Perkins had an appreciation for Treybig's business model: one that included detailed ideas for building a unique corporate culture for which Tandem became famous. One of the traditions that set Tandem apart was a weekly beer bust that included keg beer, wine, soft drinks, popcorn and unshelled peanuts. Besides being fun, the mixers encouraged employees of different groups and levels to learn what others in the company needed. Treybig was always present and accessible. Tandem was also a pioneer in granting stock options to all employees. In 1997 Tandem was acquired and it is now a server division within Hewlett Packard.

The first big system shipped to Ohio Community Libraries Cooperative. Gaye Clemson.

TANDEM

After an ad was published of two scantily clad women riding a tandem bicycle, the women in the marketing department were outraged and the result was the Incredible Hulk Contest where the male executives dressed scantily and were rated by the women of the company. Gaye Clemson.

ROLM

Rolm got its start in an old prune shack in Santa Clara. A symbol of the The Valley of Hearts Delight giving way to Silicon Valley. Pat Charles

ROLM was founded in 1969 and originally made customized and rugged computers for the US military and heavy commercial industries such as oil exploration. The company name was formed from the first letters of the founders names—Gene Richeson, Ken Oshman, Walter Loewenstern and Robert Maxfield—who were graduate students together at Stanford University. While not an original founder, Leo Chamberlain was hired on and became very much the soul of ROLM, advancing progressive workplace ideas such as GPW (Great Place to Work). Their campus on Old Ironsides Drive in Santa Clara was equipped with a swimming pool, open space park areas, a cafeteria and recreation center. In 1983 IBM partnered with, and later acquired, ROLM Corporation.

Rolm's Campus on Old Ironsides had lush landscaping, water features and plenty of outdoor seating and recreation areas. Pat Charles

Rolm's campus also featured a pool that the families of employees were encouraged to enjoy. Frequent water volleyball tournaments were hosted as a means of team bonding. Pat Charles

Rolm's lobby. Pat Charles

In 1975 Stanford engineering graduates developed the Altair, a $400 kit that included four circuit boards, and required the user to provide the monitor and keyboard. It became an instant hit with hobbyists, who bought thousands of kits in the first month. Intel.

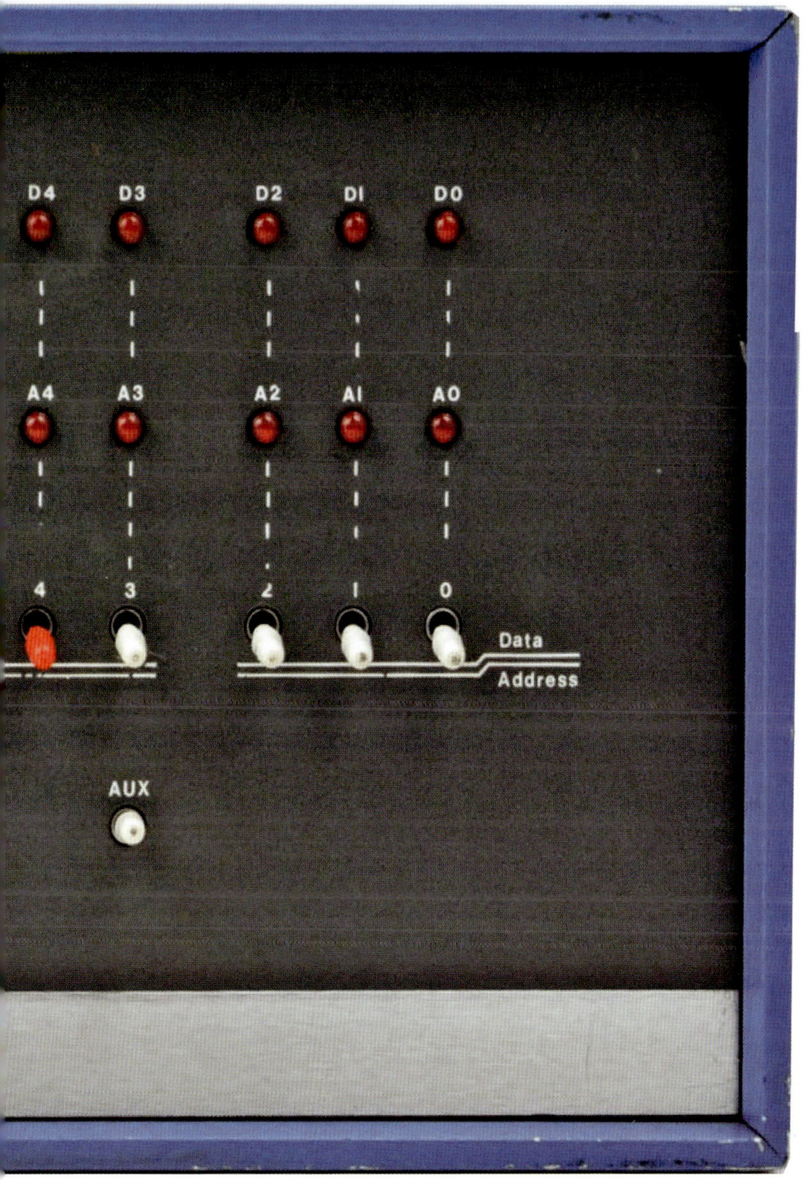

The personal computer industry was made possible by the exponential growth in processing power of semiconductor chips, starting in the late 1960s. In 1968, Hewlett Packard tried to market a programmable calculator as a personal computer. But people didn't yet understand the meaning of that terminology and sales suffered until it was re-branded as a programmable calculator. But, by 1975 the term personal computer had developed a substantial following. One of the first mainstream personal computers was the Altair 8800. This computer was developed by MIPS, a company founded by Stanford engineers that produced electronics kits for hobbyists. The Altair consisted of four circuit boards and cost $400. Because it was so affordable, it generated thousands of orders in the first month. By 1979 over half a million microcomputers were sold and the youth of the day experienced the power of the personal computer, an idea whose time had come. It was around this time that Steve Jobs and Steve Wozniak were attending Homebrew Computer Club meetings at Stanford University and developing the idea for the Apple I.

Eleven: Personal Computing

Homebrew Computer Club

Early Homebrew Computer Club member Bob Lash in his garage with a home built computer. More stories can be found online at Memoir of a Homebrew Computer Club Member by Bob Lash: http://www.bambi.net/bob/homebrew.html. Bob Lash

Famous members of the Homebrew Computer Club, Steve Wozniak and Steve Jobs. Computer History Museum.

In 1975, the Homebrew Computer Club held its first meeting in Gordon French's Menlo Park garage. The club was comprised of computer hobbyists and hackers. At the time, there was just one personal computer, the Altair 8800, based around the Intel 8080 microprocessor. While Homebrew members didn't invent the personal computer, it would be hard to overstate the impact the club had on the future of computing. Steve Wozniak, founder of Apple and early member of the Homebrew Club says of those days: *You could have called it Chips and Dips. We all had similar interests and were there to help other people, but we weren't official and we weren't formal. Our leader, Lee Felsenstein, would get up at every meeting and announce the convening of the Homebrew Computer Club which 'does not exist' and everyone would applaud happily.* Wozniak says the motto of the club was to help others and trade devices or information. In fact, he says schematics of the Apple I were passed around freely, and he would even go over to people's houses to help them build their own. As attendance grew, the meetings for the club were relocated to the auditorium at the Stanford Linear Accelerator. Afterward, many from the club reconvened at The Oasis, a bar and grill in Menlo Park which sadly closed in 2018.

Xerox PARC

Employees sit around Xerox PARC's Stanford Research Park location in this iconic photo circa 1971. Xerox PARC.

In 1970 the Xerox Corporation assembled a world class team of experts in information and physical sciences to become "The Architects of Information," and established the company's Palo Alto Research Center with the mission statement, "To Create the Office of the Future." For their new research lab, Xerox leased 14 acres of land from Stanford University and agreed to build its facility deep into the site so as not to be visible from Page Mill Road. Many Silicon Valley legends have spent time at Xerox PARC including: Steve Jobs (Founder of Apple), Andy Bechtolsheim (Founder of Sun Microsystems), Charles Geschke and John Warnock (Founders of Adobe), Pradeep Sindhu (Founder of Juniper Networks) and Eric Schmidt (previous CEO of Google).

Xerox PARC's location in Stanford Research Park. Xerox PARC.

The Alto, created by Xerox PARC's engineers in 1974, was the first workstation with a mouse and point and click functionality. Xerox PARC.

In 1971 Xerox PARC created the laser printer, which created a multi-billion dollar printing business for Xerox. Also in 1971, Xerox PARC designed Smalltalk, an object oriented programming language, which enabled computer programs to be improved without being entirely rewritten. This innovation revolutionized the software industry and influenced almost all subsequent programming systems. In 1974 Xerox PARC unveiled the Alto, the first workstation with a mouse, icons, pop-up menus, cut and paste, and overlapping windows, all of which could be controlled easily using a point and click technique. The engineers coined the phrase WYSIWYG meaning "What You See Is What You Get," to describe the graphical elements on the screen and user-friendly experience. This graphical user interface influenced the development of all subsequent computer interfaces, including the famed Macintosh.

APPLE

The Apple I was a fully assembled circuit board containing about 60+ chips. However, to make a working computer, users still had to add a case, power supply transformers, power switch, ASCII keyboard, and composite video display. Two hundred units sold in 1976.

Steve Wozniak and Steve Jobs attended Homestead High School in Cupertino where they became friends as they were both interested in electronics from an early age. After Wozniak and Jobs dropped out of college, they went to work for local technology companies: Wozniak for Hewlett Packard and Jobs for Atari. Starting in 1975, they attended meetings for the Homebrew Computer Club. In 1976 Steve Wozniak single handedly built the first microcomputer and named it Apple I. Wozniak approached his employer, Hewlett Packard, to see if they were interested in his $800 machine. They were not. So, Jobs and Wozniak raised $1,300 by selling Wozniak's HP scientific calculator and Jobs' Volkswagen van and formed Apple Computer, with the goal of marketing and selling the Apple I as a fully assembled printed circuit board.

Steve Jobs and Steve Wozniak.

Steve Jobs and Mike Markkula in 1977.

The market greeted the Apple I with open arms. With this initial success, Wozniak developed the Apple II in 1977. Once the Apple II had been developed, the partners arrived at a point where, in order to manufacture it, they needed funding. Steve Jobs approached Nolan Bushnell, his previous employer and founder of Atari, to see if he would like to make an investment in Apple. Bushnell was not interested but put Jobs in touch with Don Valentine of Sequoia Capital. Valentine, after meeting the young, unkempt Jobs asked, "Why did you send me this renegade from the human race?" Valentine was also not interested in funding Apple, but put Jobs in touch with Mike Markkula who had made millions on stock options from positions he held at Fairchild Semiconductor and Intel. Markkula was convinced of the market for the Apple II, and personal computers in general. Markkula committed $142,000 to the venture, became a one-third owner of Apple and employee number three. Markkula served as Chairman of the Board from 1985 until 1997. In 1979 Steve Jobs visited Xerox PARC and after getting a tour of the Xerox Alto, he went on to redefine personal computing modeled on the graphic user interface of the Alto. By 1980 Apple was up to several thousand employees and went public for a record $1.3 billion.

Timeline of Apple Products

The Macintosh became the first commercially successful personal computer to feature a mouse and a graphical user interface, a feature Jobs initially saw at Xerox PARC years before. Initial price tag: $3,195.

1984

Ahead of its time, Apple introduced the Newton and coined the term PDA—Personal Digital Assistant. While the Newton did not become a success, the term PDA stuck. Initial price tag: $699.

1993

1991

When first introduced, the Powerbook line was revolutionary with its built-in trackball and innovative palm rest area with the keyboard behind. All future notebooks would follow this new design. Initial price tag: $2,500.

1998

The iMac was unveiled after Jobs returned to Apple. It was considered one of their most innovative computers since the original Macintosh in 1984. It was aimed at the low-end consumer market and was designed with navigating the Internet in mind. Initial price tag: $1,299.

The introduction of the first iPod with a remarkable 5 gigabytes of storage revolutionized the music market. Initial price tag: $399.

2001

The MacBook Air was introduced as the worlds lightest computer that still maintained performance. Initial price tag: $1799.

2008

The Apple Watch syncs with an iPhone to perform functions such as calling and texting. It became the best-selling wearable device the year it was released. Initial price tag: $549

2015

2007

After much anticipation, the iPhone introduction took the market by storm. Initial price tag: $399.

2010

The most anticipated device since Steve Jobs killed the Newton, his first act as CEO when he rejoined the company. Initial price tag for the iPad: $499.

Steve Jobs

Steve Jobs' senior portrait from Homestead High School, class of 1972. Even as a child, Steve Jobs was interested in electronics. In fact, he was only 13 when he called Bill Hewlett at home and asked if he could have some spare parts for a frequency counter.

Steve Jobs and "frenemy," Bill Gates in 1985. The complex relationship between the two men began in the late 1970s when Microsoft was writing most of the software for the Apple II. Over the years, their relationship fluctuated between partners and rivals, which included both a mutual respect and cut throat competition. Supposedly, the last meeting between the two took place at Jobs' home a few months before he died.

Steve Jobs in 1980. He credits part of his success to using various seemingly unrelated experiences, to influence his products. An example he often cited was the calligraphy class he took in college, which later inspired the beautiful typography incorporated into the Macintosh.

After being ousted from Apple by John Sculley in 1985, Jobs founded NeXT Computer. Ten years later, Apple Computer acquired NeXT to use the NeXT software platform as a replacement for the dated Mac OS. After the acquisition, Jobs returned to Apple as a consultant and shortly thereafter became the de facto CEO.

In 1986 Jobs acquired the computer graphics division of Lucasfilm, which was spun off as Pixar. He was credited in Toy Story (1995) as an executive producer. He served as CEO and majority shareholder until Disney's purchase of Pixar in 2006.

Jobs died at his Palo Alto home on October 5, 2011, due to complications with pancreatic cancer. Both Apple and Microsoft flew their flags at half-staff throughout their respective headquarters and campuses. The Governor of California, Jerry Brown, declared Sunday, October 16, 2011 "Steve Jobs Day." A private memorial service for Apple employees was held on October 19, 2011, on the Apple Campus in Cupertino as seen here.

Sun Microsystems

Founders of Sun Microsystems in 1982. From left: Vinod Khosla, Bill Joy, Andy Bechtolsheim and Scott McNealy.

Sun Microsystems was founded by Stanford graduate students, Vinod Khosla, Andy Bechtolsheim, and Scott McNealy in 1982. The initial design for the Sun-1 was conceived by Andy Bechtolsheim while at Stanford. The name SUN stood for Stanford University Network. Like Steve Jobs before him, Bechtolsheim had access to the research facilities at Xerox PARC and the design of the Sun-1 was based loosely around the Xerox Alto. Shortly after the inception of Sun Microsystems, Bill Joy, a graduate student from Berkeley, joined the company and is considered one of the original founders. Sun Microsystems is also well known for the Java platform they developed in the early 1990s. Java is an object-oriented programming language which continues to be one of the world's most popular programming languages. Eric Schmidt joined Sun Microsystems in 1983 and eventually rose to become president of Sun Technology Enterprises. By the time the dot.com bubble burst in 2000, Sun had expanded too rapidly and found themselves over extended on real estate obligations and sitting inventory. Oracle acquired Sun in 2010. Andy Bechtolsheim went on to found Arista Networks and provide the initial capital for Google; Vinod Khosla founded the venture capital firm Khosla Ventures; and Bill Joy is now a partner at Kleiner Perkins Caufield & Byers.

Agnews Insane Asylum in 1896 prior to its destruction in 1906.

The Agnews clock tower, completed in 1911.

Sun's headquarters built in 1997 and now occupied by Oracle.

The campus developed by Sun Microsystems in Santa Clara has a long history. In 1885, the state built Agnews Insane Asylum. After the 1906 earthquake, the center was rebuilt and renamed Agnews State Mental Hospital. At the time, it boasted the largest population in the South San Francisco Bay Area. It was a small self-contained town with shops, a farm, a power plant, a fire department and its own train station. In 1997, Agnews campus closed and the State of California sold the land to Sun Microsystems with the condition that Sun would restore four of the historic buildings for public use.

Sun was known to have a fun-loving culture notorious for pulling elaborate pranks. One year, Eric Schmidt's office was completely relocated to a platform in the middle of a pond—complete with a working phone. Another group of pranksters converted Scott McNealy's office into a complete one-hole golf course, complete with a tee area, sand traps, water feature, ball washer and golf carts! Story and photo provided by previous Sun employee Janice Cuban.

PALM

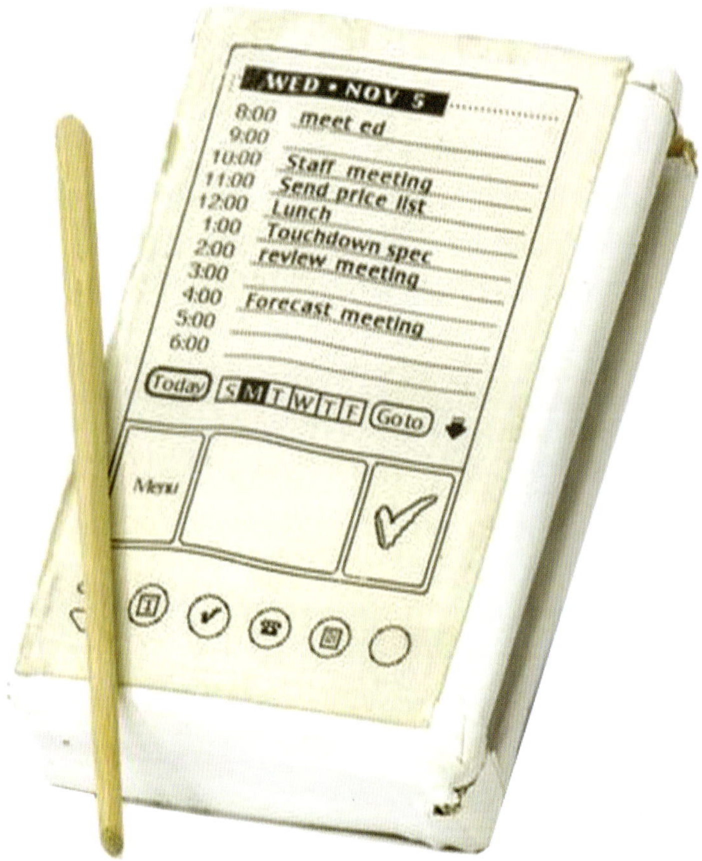

Palm founder, Jeff Hawkins created a cardboard prototype of the PalmPilot's design, using a chopstick for a stylus. The goal was to reduce the number of steps required to perform common tasks. Computer History Museum.

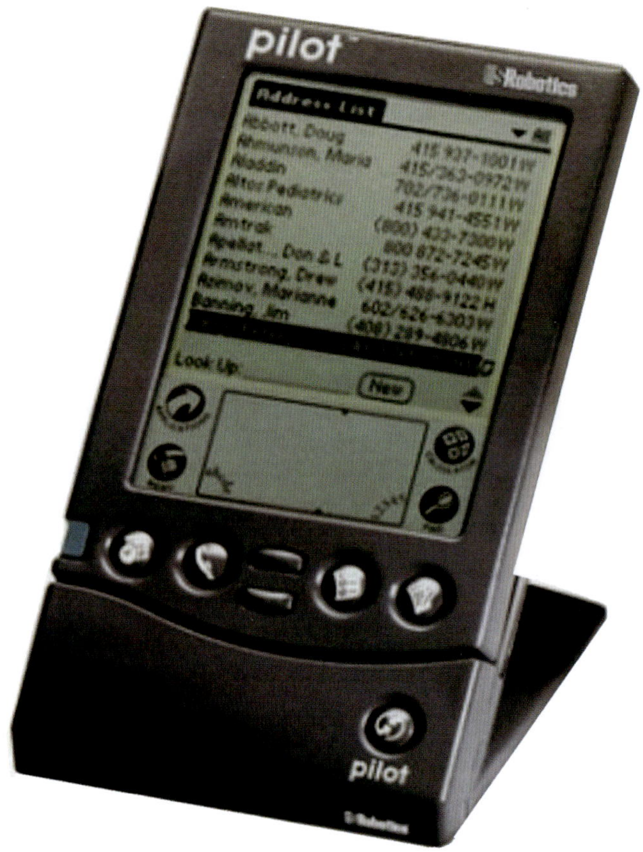

In 1996, the PalmPilot 1000 hit the market for $299. The device featured a date book, address book, to-do list, and Memo Pad which synchronized to a desktop computer using the included cradle.

Before the iPhone and even before the Blackberry, there was the PalmPilot. Jeff Hawkins founded Palm Computing in 1992. Shortly thereafter Hawkins recruited Apple veteran Donna Dubinsky to act as CEO. In 1996, the PalmPilot 1000 was introduced. It was a touchscreen personal digital assistant (PDA) that kept track of contacts, calendar items and a to-do list that synced to a desktop computer. In contrast to the Apple Newton which was released three years earlier, the PalmPilot was half the price ($299 vs $699), much smaller, and had improved handwriting recognition technology. The PalmPilot is credited with popularizing the PDA and proving there was a market for a hand-held computer. In 1997, Palm became a 3Com subsidiary. The merger did not sit well with Hawkins, Dubinsky and Ed Colligan, then the head of marketing at Palm. The three executives left to found Handspring.

The Trēo line of smartphones offered the functionality of a PDA which allowed a user to dial directly from a list of contacts, and send email messages. Later models included a built-in camera and navigation.

In 2002, Handspring introduced the Trēo smartphone line. These devices combined a cell phone with the functionality of a PDA and the ability to browse the Internet. Through many iterations of corporate spin-offs, IPOs and mergers, a then-independent Palm, Inc. merged with Handspring in 2003. The Trēo 600 was the first device offered under the new palmOne brand. In 2010, Palm was purchased by Hewlett-Packard but a year later, HP announced that it would end production of Palm devices which marked the end of the Palm brand after 19 years.

Palm's first employees as seen here in 1992. Donna Dubinsky

From left to right: Ed Colligan, Donna Dubinsky and Palm Founder, Jeff Hawkins. This team has been credited with popularizing the personal digital assistant, or PDA. Computer History Museum

Computer History Museum

Located in Mountain View, the Computer History Museum traces the advancement of computing from early tools like the abacus through today's cutting-edge applications like artificial intelligence. The Museum dates back to 1975 when Digital Equipment Corporation, then headquartered in Massachusetts, converted a lobby coat closet into a small museum. After occupying a few locations on the East Coast, the Computer Museum History Center was established at Moffett Field in 1996. In 2003, CHM opened the doors to its current building on Shoreline Boulevard which was previously occupied by Silicon Graphics. The facility is over 100,000 square feet and features over 90,000 objects, photographs, and films. The Museum prides itself on being home to the largest collection of computing artifacts in the world. On display is a compilation of hardware, software, prototypes, manuals, business plans, lab papers, marketing materials, and oral histories. Their collection includes many rare or one-of-a-kind objects including a Cray-1 supercomputer, the 1969 Neiman Marcus Kitchen Computer, and an Apple I. In addition to rotating exhibits, CHM has two major exhibitions—Revolution: The First 2000 Years of Computing (opened in 2011) and Make Software: Change the World! (Opened in 2017). A few of the items on display at the museum are featured on the opposite page.

NEIMAN MARCUS KITCHEN COMPUTER

Why would anyone want a computer at home? To store recipes, of course! In 1969, Neiman Marcus debuted the Kitchen Computer. For $10,600 you got the computer, a cookbook, an apron, and a two-week course. The futuristic console featured a built-in cutting board and a Honeywell 316 minicomputer. One problem was that the interface only displayed recipe data in binary lights. Does 0011101000111001 mean broccoli? Or carrots? Not surprisingly, none were sold. CHM

UTAH TEAPOT

The Utah teapot refers to a 3D model that is now a standard reference object for computer graphics. The graphic was created in 1975 by Martin Newell, an early computer graphics researcher at the University of Utah. The teapot has now become a running joke in the industry and references can be found in animations like Toy Story and The Simpsons. Ironically, now physical copies of the teapot are sold as keepsakes thanks to the advent of 3-D printing. CHM

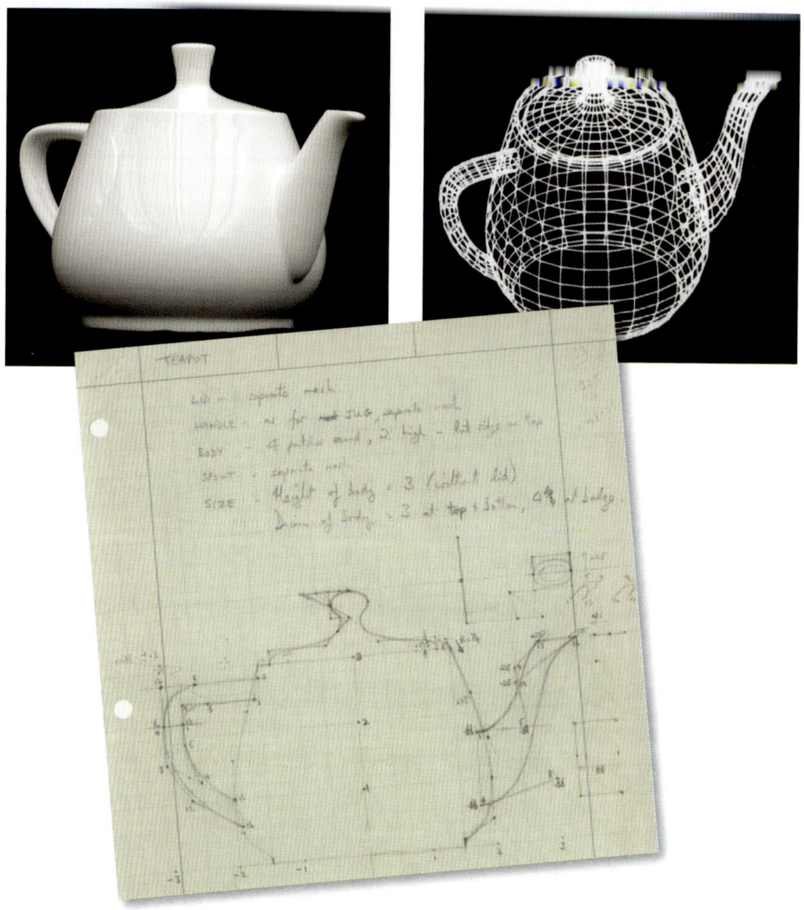

EDISON PHONOGRAPH DOLL

Thanks to MP3s, many people today take for granted how music is now instantly available and easily shared. However, long before MP3s, an early portable music player was a doll! Thomas Edison invented the phonograph doll in the 1870s, and variations were made for many decades. Like early home phonographs, music came on wax cylinders. CHM

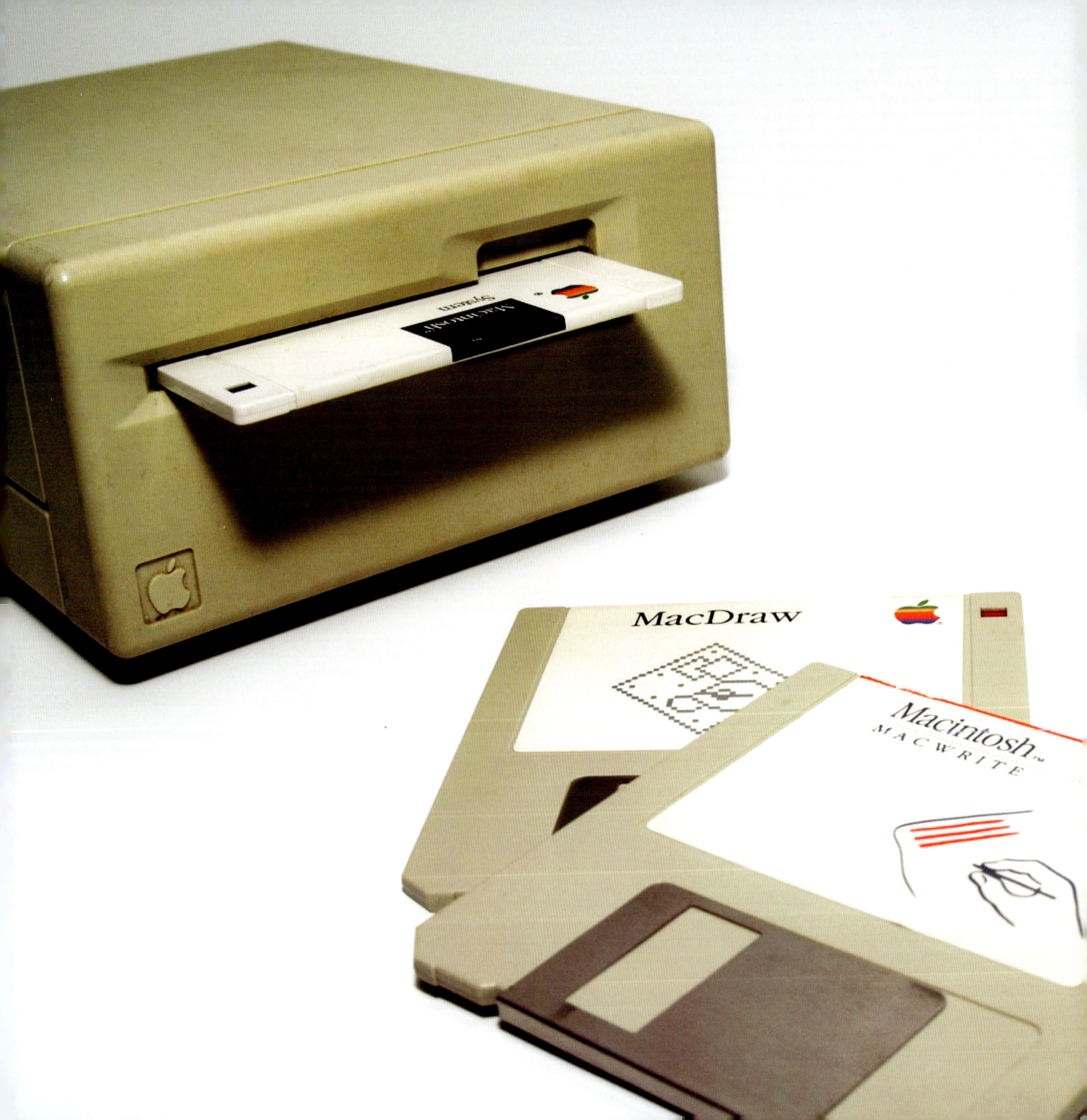
MacDraw
Macintosh
MACWRITE

MacWrite and MacPaint were early software programs developed by Apple Computer. Originally released in 1984, the pair of programs were delivered on floppy disks and sold for $195. Wikimedia

Software is essentially a set of instructions that tell a computer what to do. The first generation of software for mainframe computers had instructions written in binary code. A binary digit can have only one of two values, most commonly represented as either a 0 or 1 but can also be understood to have the values true/false, yes/no, +/– , on/off, etc. A binary string of only eight digits has 256 possible value combinations and therefore can hold a wide range of information. Today, more complex software programs are created using a programming language. Many consider Fortran to be the first programing language. Invented in 1957 by a team at IBM led by John Backus, Fortran is still in use today. The software industry exploded with the rise of the personal computer in the 1970s. In 1971 Xerox PARC designed Smalltalk, an object-oriented programming language which enabled computer programs to be improved without being entirely rewritten. This innovation revolutionized the software industry and influenced almost all later programming systems. More recently software has evolved to be hosted online instead of being loaded directly onto the machine, spawning the name software-as-a-service, or SaaS.

Microsoft has long been known as the largest software company in the world. Microsoft has also been known for not playing well with others. This is in sharp contrast to Silicon Valley which believes it is mutually beneficial to embrace an open-source community. Most Silicon Valley companies are at least on frenemy status with their adversaries. The fact that Microsoft is not located in Silicon Valley has historically perhaps been an advantage—it was able to operate a near monopoly and not play by the Valley's rules. However, in 2012 the software behemoth posted their first quarterly loss in their 26 year history and some wonder if having an isolationist mentality is partly to blame.

Twelve: Software

ORACLE

◀ *Aerial shot of Oracle's campus in Redwood Shores.*
Rehman Abubakr via Wikimedia

▼ *Oracle founders circa 1978. Left to right: Ed Oates, Bruce Scott (early employee), Bob Miner and Larry Ellison.*
Computer History Museum

In 1977 Larry Ellison co-founded Oracle with Bob Miner and Ed Oates to pursue the growing market of database management. Prior to founding Oracle, Ellison worked at Ampex where he was tasked with designing a database for the CIA. Around this time, he read a paper by IBM engineer Edgar F. Codd called *A Relational Model of Data for Large Shared Data Banks*. IBM was slow to pick up on Codd's theories of data arrangement and Ellison commercialized it first. In 1979 the Oracle 2 (there was no Oracle 1) was introduced. It was the first RDBMS (Relational DataBase Management System) to use SQL (Structured Query Language). SQL allows a user to access and manipulate databases. While Oracle was the first to commercialize SQL, it became an international standard in 1987. Today, Oracle is still in the same business they were in when they got started: database management. Oracle is the second largest software company after Microsoft. Opposed to Microsoft who has tens of millions of customers, Oracle has only 400,000 customers. Even more crazy is that it derives 50% of its revenue from the top 1% of its clients. Oracle has a few competitors, but none offer a product as versatile, reliable, or secure. Oracle knows this and is able to charge its customers accordingly.

Larry Ellison

Larry Ellison circa 2009. Oracle

Team Oracle sailboats in the San Francisco Bay circa 2013.
Don Ramey Logan via Wikimedia

Even if you don't know exactly what Oracle does, you probably know of its charismatic co-founder, Larry Ellison. Ellison served as the CEO of the company from 1977 until 2014 when he transitioned to Chief Technology Officer. Today, he continues to serve as Chairman of the Board. As of 2018 Forbes magazine pegged him as the eighth-wealthiest person in the world. A few factoids:

* Ellison was born in New York City to an unwed mother whom he did not see again until he was 48. He was raised by his aunt and uncle who were Russian immigrants and chose the last name Ellison as a tribute to Ellis Island.
* In 1999, Ellison's protégé Marc Benioff left Oracle to found Salesforce. Ellison invested $2 million into his friend's new venture. However, when Benioff discovered that Oracle was simultaneously working on a product to compete with Salesforce, he tried to get his mentor to quit the board. Instead, Ellison forced Benioff to fire him which allowed Ellison to keep his shares in Salesforce.
* Ellison also sails yachts competitively. Ellison financed Oracle Team USA to compete in the 2003 America's Cup race. His team competed again in the 2007 event before winning the race in 2010. The team won the race again in 2013. The boat that led the team to victory in 2010 was flown via helicopter to Redwood Shores and now resides in the pond outside Oracle's headquarters.

Adobe

Adobe founders, John Warnock and Charles Geschke. Adobe

Adobe's first office in Mountain View on Marine Way, circa 1982. Adobe

In 1982 John Warnock and Charles Geschke left Xerox PARC to found Adobe. The company name comes from the Adobe Creek in Los Altos, which ran behind both founders' homes. Their first product was the PostScript page description language, which described to a printer the appearance of a printed page. Adobe's Postscript language went on the market in 1984. Around the same time, Steve Jobs approached Adobe and urged them to adapt PostScript to be used as the language driving laser printers. In March 1985 the Apple LaserWriter was the first printer to ship with PostScript, sparking the desktop publishing revolution in the mid-1980s. The next move for Adobe was a to enter the consumer software market with Adobe Illustrator, a vector-based drawing program for the Apple Macintosh. In 1989 Adobe introduced what was to become its flagship product, an image-editing program for the Macintosh called Photoshop. In 1993 Adobe introduced PDF, the Portable Document Format: and Adobe Acrobat/Reader software, which became an international standard. As of 2018 Adobe had around 18,000 employees, about 40% of whom work in San Jose.

Adobe PostScript

PostScript Language

1984

Adobe Photoshop

1989

Adobe PageMaker

1994

1986

Adobe Illustrator

1993

Adobe PDF

1999

Adobe InDesign

Adobe now occupies three high rise buildings in downtown San Jose totaling around 1 million square feet. The buildings are known for their green design and in 2006 the towers were awarded The Leadership in Energy and Environmental Design (LEED) Platinum Certification by the United States Green Building Council. Adobe

Symantec

Gary Hendrix at original Symantec headquarters circa 1982. Gary Hendrix

1987 photo of early Symantec employees. Founder Gary Hendrix is sitting on the floor in a pink shirt and former long time CEO Gordon Eubanks is in the chair with a blue blazer. Computer History Museum

Symantec was founded in 1982 by Gary Hendrix with a grant from the National Science Foundation and additional funding from Kleiner Perkins. Prior to founding Symantec, Hendrix worked at the Stanford Research Institute (SRI) on a project for the Navy. There, he was tasked with developing a software which would allow users to quickly search large databases by questioning a computer in plain English (as opposed to artificial or programming languages). This research led Hendrix to see there was a larger application for natural language programming. This lent to the company name Symantec: a combination of the words "syntax, semantics, and technology". In 1984 George Eubanks joined Symantec. He would serve as president and CEO, guiding the company into the software utility and anti-virus business. In 1985 Symantec launched its first widely successful product, a database system built for the IBM PC, which allowed users without advanced computer training to query databases in plain English. It was called Q&A short for Question & Answer. Symantec went public in 1989.

One of Symantec's most well-known products over the years: Norton AntiVirus.

John Thompson, Symantec's CEO from 1999–2009 who now serves as the Chairman of the Board for Microsoft. NASDAQ

By the late eighties Symantec was facing growing competition. The company responded by turning their focus to a new and undeveloped market segment—anti-virus software. The company strengthened their position in this emerging market through a number of acquisitions including Peter Norton Computing in 1990. In 1999 John Thompson became CEO of Symantec, a post he held for the next ten years. Under his direction Symantec focused squarely on security, dropping many of the original product lines. This completed Symantec's evolution from artificial intelligence to cyber intelligence company. CEO Greg Clark joined the company in 2016 after Symantec's acquisition of Blue Coat. The Chairman of the Board is Dan Schulman who also serves as PayPal CEO. Today Symantec has nearly 11,000 employees and continues to dominate the cybersecurity industry.

Symantec's current CEO Greg Clark.

INTUIT

Scott Cook, Founder of Intuit, with an early copy of the Quicken Software in the late 1980's. Intuit

Below: A 1984 copy of Quicken Software in front of an IBM PC. On the screen is a digital check to be filled out by the user. Intuit

Intuit was conceived by Scott Cook in 1983. His earlier work at Procter & Gamble provided insight into the future of personal computers. He was inspired to create a platform for personal accounting on the home computer. Armed with an idea, Cook connected with Tom Proulx through Stanford University and initially operated out of a single room on University Avenue in Palo Alto. The first version of Intuit's original flagship product, Quicken, was coded by Tom Proulx for the IBM PC and the Apple II. By the early nineties, the company faced major competition, especially with the release of Microsoft Money in 1991. Intuit innovated their product line and engaged John Doerr of Kleiner Perkins to help them grow the company. Doerr remained on the board through 2007. In 1993, Intuit went public at which point Cook decided a new CEO was needed to propel the company to the next level. Bill Campbell was tapped for the job. Campbell has since been nicknamed "the coach of Silicon Valley Stars" due to his advisory role to the likes of Steve Jobs and Larry Page. Campbell became an integral part of Intuit and assumed the position of Chairman of the Board in which he remained until he passed away in 2016.

Quicken	intuit quickbooks	TurboTax	intuit QuickBooks Online	intuit mint
1984	*1992*	*Acquired 1993*	*1995*	*Acquired 2009*

Bill Campbell and Scott Cook at the Old Pro in Palo Alto. Intuit

Brad Smith and Bill Campbell speak to Intuit employees during a company-wide event in 2012. Intuit

Brad Smith joined Intuit in 2003 as a general manger. He served as CEO from 2008 through the end of 2018. By 2013 Intuit had a customer base of 60 million users across its product line as it began to transform from a desktop software company to a cloud-based technology company. In 2016, Intuit sold its original product, Quicken. Today, Intuit continues to innovate with their flagship products—QuickBooks, TurboTax, Mint & ProConnect. Intuit's Mountain View campus is comprised of 11 buildings where they employ over 7,900 people. They have a market cap of more than $20 billion and have been voted one of America's "Best Companies to Work For" by Fortune magazine for 16 consecutive years.

"The van where the Internet was born". SRI.

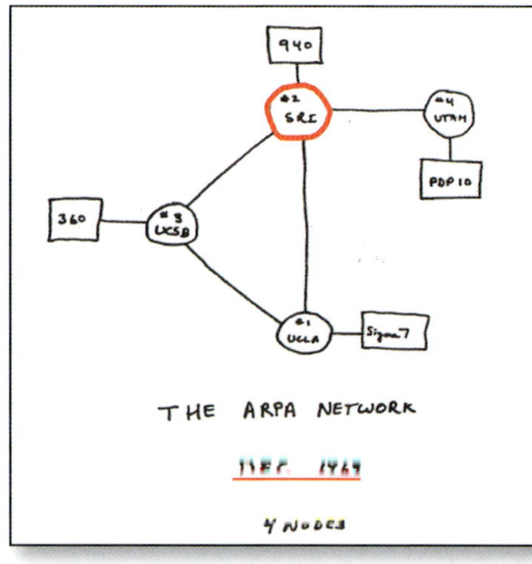

The hand sketched design of the ARPANET, which transmitted the first message over an electronic computer network in 1977. SRI.

In 1958 the Russians launched Sputnik, the first earth orbiting satellite. In response, the U.S. Government created a new task force, the Defense Advanced Research Projects Agency or DARPA, whose mission was to expand the frontiers of technology and science as a matter of national defense. One of DARPA's original projects was the ARPANET which was the precursor to the Internet we know today. The first true Internet transmission occurred in November 1977 when SRI originated the first connection between three disparate networks. However, it wasn't until the mid 1990s that the concept of the Internet really took shape and began its rapid expansion. During the early years of the Internet, most people connected over a phone line with a dial-up 56K modem. To put this into context, that is almost two thousand times slower than connections today. In 1995 about 10% of adults in the U.S. were going online. By 2011 that figure had jumped to 78% of adults and 95% of teenagers.

Thirteen: The Internet

Stanford Research Institute

Doug Engelbart at SRI, showcasing the first computer mouse which he designed. SRI.

In 1946 the trustees of Stanford University established the Stanford Research Institute as a nonprofit organization to serve as a center of innovation to support economic development in the region. SRI International, as it is now known, formally separated from Stanford University in 1970. To this day, it remains one of the largest contract research institutes in the world performing research and development for government agencies, commercial businesses, and private foundations. It also licenses its technologies, forms strategic partnerships, and creates spin-off companies. One famous alumni is Doug Engelbart as seen above. He was the main innovator of modern computer-human interface elements including: bit-mapped displays, collaboration software, hypertext, and the computer mouse. He was awarded the National Medal of Technology and Innovation in 2000. He passed away in 2013.

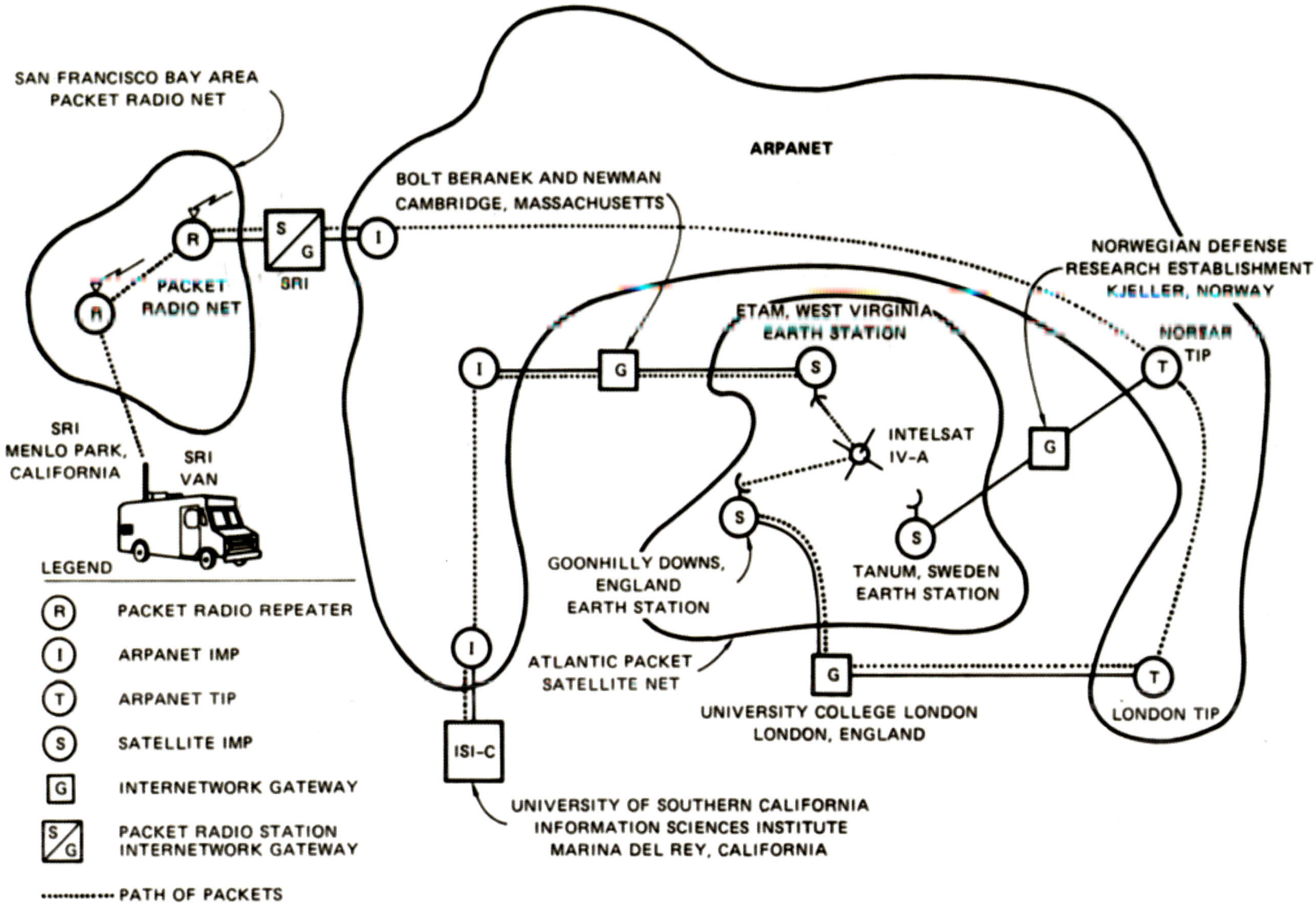

A schematic of the 1977 connection between three disparate networks. SRI.

Stanford Research Institute was integral in launching the Internet. In October 1969 the world's first electronic computer network, ARPANET, was established between nodes at UCLA and Douglas Engelbart's lab at SRI. The first message on the ARPANET was sent by a UCLA student with the message text 'login'. The l and the o letters were transmitted, but the system then crashed. Hence, the literal first message sent over the ARPANET was lo. The first true Internet transmission occurred in November 1977, when SRI originated the first connection between three disparate networks. Data flowed seamlessly through a mobile van between SRI in Menlo Park, California and the University of Southern California in Los Angeles via London, England, across three types of networks: packet radio, satellite, and the ARPANET.

Cisco Systems

Cisco founders, Sandy Lerner & Len Bosack. Cisco.

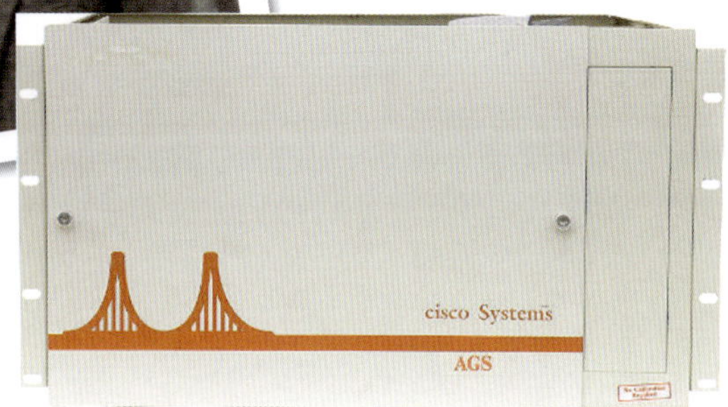

Cisco's AGS Router introduced in 1986.
Mark Richards / Computer History Museum.

Logo circa 1985.

Logo circa 2002.

Logo, present day.

Cisco Systems was founded in 1984 by then couple Len Bosack & Sandy Lerner. Bosack came to Stanford University to pursue a graduate degree in computer science in 1979. While at Stanford, he met Lerner who was working as the Director of Computer Facilities for the Graduate School of Business. The pair married in 1980 and started Cisco Systems out of their home in Atherton in 1984. The company provided the first commercially viable multi-protocol router that allowed disparate computer "networks" to "connect" to one another and got its name from the city of San Francisco with a logo that symbolizes the Golden Gate Bridge. In 1987, the famed venture capitalist Don Valentine agreed to provide the young company with a $2 million investment in exchange for stock ownership and control of the company. One of Valentine's first moves was to remove the founders from their management positions and within a year, had installed John Morgridge as CEO in 1988. As the popularity of the Internet grew, the demand for networking devices also grew. By 1990, the company had grown to 250 employees with $69 million in revenue and decided to go public. Cisco's IPO is considered one of the greatest of all time due to its explosive growth in subsequent years. According to Investopedia, a $1,000 investment during Cisco's IPO would be worth over $450,000 today.

Top: John Chambers at the Cisco Networkers event in 1996. Cisco had recently shipped its millionth product. Cisco
Bottom: Participants were given railroad engineering hats at the first Networkers meeting in 1989 to illustrate the parallels between building the Internet and constructing the railroads during the previous century. Zany hats continue to be a hallmark of the Networkers conference today.
Mark Richards / Computer History Museum.

An example of the Internet of Things – Smart Lighting – has the potential to reduce energy costs by 90 percent. These LED lights automatically brighten and dim as needed, provide feedback on energy consumption and feature a camera which monitors traffic, parking, safety and security. Cisco.

Cisco launched TelePresence in 2006 – a next-generation video technology which enables people to engage and innovate anywhere, on any device. From left to right, Cisco CEO John Chambers provides a demonstration of the solution with President George W. Bush and Governor Arnold Schwarzenegger in April 2006. Cisco.

John Chambers joined the company in 1991 and was appointed CEO in 1995, a position he would hold for 20 years. He helped grow the company from $70 million when he assumed the role of CEO, to $47 billion in fiscal year 2015 when he stepped down. As of 2017, Chambers became Chairman Emeritus. His leadership was instrumental in helping Cisco achieve market dominance by placing importance on continual R&D innovation as well as strategic company acquisitions. Chambers has received numerous awards for his leadership over his 20 years at the helm of Cisco, including being named the number two "Best-Performing CEOs in the World" in 2015 from Harvard Business Review, one of Time Magazine's "100 Most Influential People" and was the recipient of the Edison Achievement Award for Innovation. 1990-2010 marked the Information Age. We are now in the Digital Age and Cisco is leading the way with the Internet of Things – where the physical objects of our day to day lives, including devices, vehicles, and buildings are embedded with sensors, software and network connectivity. It's about getting the right information, to the right place, at the right time to intelligently collect, exchange and analyze data. Today, Cisco has more than 70,000 employees located in more than 165 countries.

Juniper Networks

Juniper Networks Founder Pradeep Sindhu with former long time CEO Scott Kriens who has served as the Chairman of Board since 1996. *Juniper Networks.*

First day of trading on the New York Stock Exchange in 1999 where the stock price rose 191% and set a record for first day trading in the technology sector. Juniper Networks.

In the early days of the World Wide Web, data traveled across the Internet using routers initially designed to handle phone calls. This method transmitted the entire collection of data at once, which was sent over one dedicated path. In 1995, Pradeep Sindhu was working at Xerox Parc and while on vacation he conceived of a way to better handle Internet traffic: packet-based routers. Packet-based routers break up the data into smaller units allowing each "packet" to find the most efficient route to its ultimate destination. With this, Juniper Networks was born. Sindhu founded the company with $2 million in seed funding, followed shortly by an infusion of $12 million. Within the first year of operation, Scott Kriens was appointed CEO to manage the business, while founder Sindhu became the Chief Technology Officer. In 2007, Juniper Networks raised an additional $40 million with investments from four of the largest telecommunications equipment manufacturers: Siemens, Ericsson, Nortel, and 3Com.

Juniper Networks was founded in Sunnyvale in 1996 and has stayed true to its roots, now occupying several buildings in the Moffett Park neighborhood. Juniper Networks.

Juniper Networks went public in 1999. The first day on the NYSE, the company's stock rose 191% and is considered one of the most successful initial public offerings in history. Within a year, the company's stock grew five-fold propelled by their rapid growth (2000 saw 600% revenue growth) and gains in market share (Juniper Networks had 1/3 of the market for high-end core routers by 2001). Around this same time, market analysts were keen to speculate on the competition between Cisco and Juniper Networks and often called the rivalry as the "David vs. Goliath" of Silicon Valley. The company has continued to thrive and innovate and is now focused on creating solutions, products and services in the cloud era to transform the way the world connects, works and lives. As of 2018, Juniper Networks has 92 offices in 43 countries with 9,500 employees and over $5 billion in annual revenue.

NETSCAPE

Netscape's Jim Clark, left, and Marc Andreessen pose outside of their offices in Mountain View.

Marc Andreessen was raised in the Midwest. While at and the University of Illinois, he worked at the National Center for Supercomputing Applications (NCSA) which was located on campus. There he became familiar with standards for the World Wide Web. He partnered up with Eric Bina to work on creating a user-friendly web browser containing a graphical user interface. The resulting code was the Mosaic Web Browser. After Andreessen graduated in 1993, he moved to California and met with Jim Clark, the recently departed founder of Silicon Graphics. Clark believed that the Mosaic browser had great commercial possibilities and suggested starting an Internet software company. Clark helped secure funding from Kleiner Perkins Caufield & Byers and soon Mosaic Communications Corporation was in business.

The original Netscape Internet browser.

The company soon changed its name to Netscape Communications, and its flagship Web browser was the Netscape Navigator. Netscape made a very successful IPO in 1995. The stock was initially priced at $14 per share and closed at $58.25 that same day, creating a market value of $2.9 billion. The runaway success of Netscape's IPO turned Andreessen into the poster-boy wunderkind of the Internet generation: twenty-something, high-tech, ambitious, and worth millions (or billions) of dollars practically overnight. Netscape's success attracted the attention of Microsoft, which recognized the Web's potential and wanted to put itself at the forefront of the rising Internet revolution. Microsoft licensed the Mosaic source code from the University of Illinois, and turned it into Internet Explorer. The resulting battle between the two companies became known as the Browser Wars. Netscape was acquired in 1999 for $4.2 billion by AOL, which made Andreessen its Chief Technology Officer. Marc Andreessen went on to found the venture capital firm Andreessen Horowitz in 2009 with partner Ben Horowitz, and the pair have since invested in companies such as Twitter, Groupon, Skype and Zynga.

YAHOO!

David Filo and Jerry Yang in 1997 with a fish prop in Yahoo!'s early office located at 3400 Central Expressway in Santa Clara.

Jerry Yang and David Filo in front of their Yahoo logo.

Jerry Yang and David Filo were electrical engineering graduate students at Stanford University. In 1994, while trying to organize their personal interests on the Internet, they created a website they called, "David and Jerry's Guide to the World Wide Web." It was a directory of other websites, organized in a hierarchy with categories and subcategories. They soon found that they had a loyal following of people who also wanted a single place to find useful Web sites. Later that year, they renamed their site to "Yahoo!" which is an acronym for "Yet Another Hierarchical Officious Oracle." The founders say they also liked the name based on the definition of a yahoo: "rude, unsophisticated, uncouth." Yahoo!'s fan base continued to grow and within six months their site was receiving 1,000,000 hits per day. With such an enthusiastic reception, the pair incorporated the business and raised a $2 million round of funding from Sequoia Capital in 1995. The site's popularity grew exponentially, and with only 49 employees they went public in 1996. The company was hit hard by the dot.com bust and has gone through a number of management changes during the last decade. In 2012, Yahoo! appointed former Google executive Marissa Mayer, to the post of CEO. In 2016, Verizon agreed to purchase the core business of Yahoo for $4.83 billion.

Marissa Mayer

Marissa Mayer in 2009. *Eirik Solheim via Wikimedia.*

Marissa Mayer grew up in Wisconsin. She attended Stanford University and interned at SRI International. Out of the 14 job offers she received after graduation, she decided to accept the offer from Google to become their twentieth employee—and first female engineer. At Google, Mayer held key roles in almost all aspects of the company. She was instrumental in helping design the clean, no-nonsense home search page. She left Google in 2012 to become CEO of Yahoo! in an attempt to save the struggling company. Mayer's appointment to CEO was met with excitement and optimism. One of her first action items was to present the CEO Challenge—an invitation to all Yahoo! employees to submit their ideas on how to save the company. She received over 700 entries. For all her efforts, she wasn't able to reverse the slowing growth nor adequately address declining employee morale. Mayer resigned as CEO of Yahoo! in 2017 when the sale to Verizon was completed. In 2018, Mayer co-founded Lumi Labs, an incubator focused on consumer media and artificial intelligence. The tech incubator is housed at 165 University Avenue in downtown Palo Alto. The building has good Silicon Valley juju—Google and Paypal both got their start in this building; YouTube provides this address as the example when setting the location of an uploaded video.

Famous Yahoo! Alumni

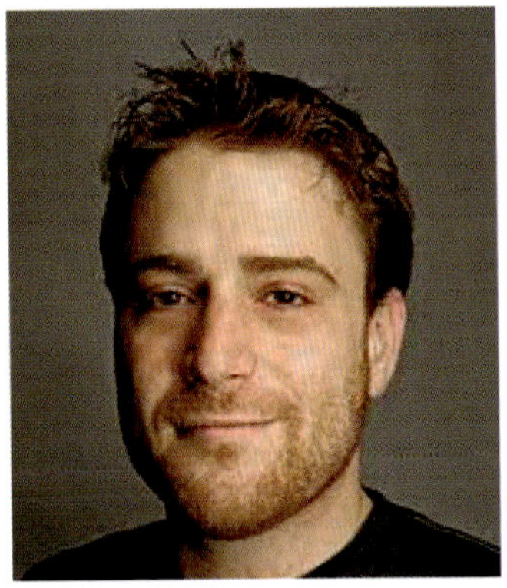

Stewart Butterfield, Slack CEO

Stewart Butterfield joined Yahoo! when it bought his photo-sharing app, Flickr, in 2005. Butterfield left Yahoo! in 2008. In 2014 Butterfield released Slack, a set of team collaboration tools. The name is an acronym for *Searchable Log of All Conversation and Knowledge*. As of 2018, it is valued at over $7 billion and considered one of the fastest-growing startups ever.

Dan Rosensweig, Chegg CEO

Dan Rosensweig joined Yahoo! as COO in 2002, a post he held for five years. In February 2010, Rosensweig joined Chegg as President and CEO. Chegg originally specialized in online textbook rentals (both in physical and digital formats) although has expanded into homework help, tutoring, scholarships and internship matching.

Blake Irving, GoDaddy CEO

Blake Irving joined Yahoo! in 2010 and served as Chief Product Officer for 2 years. Irving was a part of the team at Xerox Parc in the days of WYSIWYG. Irving spent 15 years at Microsoft working on MSN Messenger and Hotmail. Irving gave up his semi-retirement to become GoDaddy's CEO in 2013. Under his leadership, the company ditched the sexually provocative commercials which ultimately helped the bottom line.

Jeff Weiner, LinkedIn CEO

Jeff Weiner joined Yahoo in 2001. He left in 2008 to join LinkedIn as the Interim President. Weiner became LinkedIn's CEO in 2009 and played an instrumental role in Microsoft's acquisition of LinkedIn in 2016. He currently also sits on Intuit's Board of Directors.

Brian Acton and Jan Koum, WhatsApp Co-founders

Brian Acton and Jan Koum joined Yahoo! in 1996 and spent nearly 9 years at the company. In 2007, the pair left Yahoo. They applied to work at Facebook but weren't accepted. Ironically, the pair went on to found WhatsApp two years later which was acquired by Facebook in 2014 for $19 billion.

Dave Goldberg, ex-Survey Monkey CEO

Dave Goldberg joined Yahoo in 2001 when they acquired his online music company, Launch Media, in 2001. He spent nearly 6 years at Yahoo! Dave Goldberg joined Survey Monkey in 2009 and is credited with growing it from a 12-person startup to a 550-person enterprise. He served as CEO until his sudden death in 2015 while on vacation with his wife, Facebook executive Sheryl Sandberg.

Rich Riley, Shazam CEO

Rich Riley joined Yahoo! in 1999 when they acquired his company, Log-me-on.com. He then spent 13 years at Yahoo! until he left to join music discovery app Shazam as its CEO. Apple acquired Shazam in late 2018.

Google

Sergey Brin and Larry Page started Google in Susan Wojcicki's garage. To celebrate the companies twentieth anniversary, Google released a new street view tour so you can take a stroll through the garage as it was 20 years ago. The house on Santa Margarita Avenue in Menlo Park is now owned by the tech giant. They have recreated the scene as seen here complete with many CRT monitors, cables, a ping pong table, and a keyboard for music breaks. Google.

Larry Page and Sergey Brin met while studying for a graduate degree in computer science at Stanford University. In 1995, the pair began collaborating on an Internet search engine they called BackRub. In the beginning, they operated their search engine on Stanford servers but by 1997 they decided to commercialize it. They named their new company Google—a play on the word googol, a mathematical term for the number one followed by one hundred zeros. The term reflects their mission to organize a seemingly infinite amount of information on the web. In 1998, Sun co-founder Andy Bechtolsheim gave Sergey and Larry $100,000 to get the company off the ground. The company first started in Susan Wojcicki's garage in Menlo Park. Only a few months after incorporating, PC Magazine named Google the search engine of choice. In 1999 with only eight employees, Google left the garage for their first corporate location: 165 University Avenue in Palo Alto. Within a few months, the company announced a $25 million round of funding from Sequoia Capital and Kleiner Perkins, as well as the need to relocate the corporate location again due to their expanding workforce. With over 35 employees, Google moved to the Shoreline district of Mountain View, with their first building located at 2400 E. Bayshore Road. With this move they hired their first corporate chef, Charlie Ayers, whose initial claim to fame was catering for the Grateful Dead. By the year 2000 Google had returned over a billion search results and became the world's largest search engine.

In 1999, Google left the garage for their first corporate location—165 University Avenue in downtown Palo Alto. Sergey Brin, Larry Page and Marissa Mayer can be seen standing together in the back right. Susan Wojcicki is all the way on the left. Google.

Google Culture

Google is known for its great corporate culture. The tech company aims to make their expansive Silicon Valley campus a fun place to be, where it doesn't feel like work. Perks include free meals, massages, dry-cleaning, and the ability to bring your dog to work. Its a culture many try to emulate.

Communal bikes are provided by Google for employees to travel around the campus.

Google's iconic Shoreline campus in Mountain View.

Aerial view of the Mountain View campus where most buildings are powered with solar panels.

LARRY PAGE was born in Michigan in 1973. Both of his parents were early pioneers in the fields of computer science and he began to play around with computers at an early age. While working on his PhD at Stanford University, he set out to better understand the World Wide Web and the links that connect one web page to another. Page was inspired by the similarities shared between the web and how academics build a research paper—namely the importance of citations. The more a paper is cited (or a page is backlinked), the more intrinsic value it has. Google's search algorithms are a result of his early research project named BackRub. Page served as Google's first CEO until 2001. He re-assumed the role in 2011, now serving as CEO of Alphabet, Google's parent company. As of late 2018, Page was estimated to be the ninth richest person in the world.

SERGEY BRIN was born in the Soviet Union in 1973. He immigrated to the US when he was 6 years old. He earned his bachelor's degree at the University of Maryland and then attended Stanford University to pursue a PhD. While in his second year at Stanford, he volunteered to be a guide and mentor to potential first-year students. It was through this position that he met Larry Page. Brin is gregarious by nature; Page is more reserved. The pair initially clashed although soon became fast friends. Shortly thereafter, Brin joined Page to work on the BackRub project. Today Brin is the President of Google's parent company Alphabet and is estimated to be the thirteenth richest person in the world.

ERIC SCHMIDT was born in Virginia in 1955. Early in his career he worked for Xerox Parc, Sun Microsystems and Novell. In March 2001, he joined Google as the Chairman of the Board and a few months later took over the roll of CEO. Schmidt shared responsibility for Google's daily operations with founders Page and Brin until he stepped down from the role of CEO in 2011 and the role of Chairman in 2017. His first day on the job he was taken aback to learn he would have to share an office with an engineer. He ended up befriending the engineer and the pair chose to share an office for years. It serves as an example of the culture at Google—insubordination isn't necessarily frowned upon and anyone with a good idea has the ability to be heard.

SUNDAR PICHAI was born in India in 1972. After college in India, he immigrated to the US to receive his MS from Stanford and then his MBA from Wharton. He worked at Applied Materials and McKinsey & Co. before joining Google in 2004. He was promoted to CEO of Google in 2015, as part of the restructuring process that made Alphabet Inc. into Google's parent company.

Alphabet, Inc.

In 2015, Google restructured and created Alphabet to become the parent company for Google and its subsidiaries. The two founders of Google assumed executive roles in the new company—Larry Page is CEO and Sergey Brin is President. Google remains the largest subsidiary under this new parent company but there are a number of other lesser known companies, including:

X (formerly Google X): A semi-secret research lab founded in 2010. The mission is to invent and launch moonshot technologies with the aim to radically transform the world. Many successful X experiments are now formal companies under the Alphabet umbrella. Past and present projects under X include Google Glass, delivery drones, Internet balloons, and kites designed to harness wind energy.

Google Ventures: Launched as Google Ventures in 2009, GV is the venture capital arm of Alphabet, Inc. They have invested in more than 300 companies as of 2018. They have investments in companies ranging from artificial intelligence, robotics, transportation and cyber security to agriculture and healthcare. A few notable investments include ride sharing company Uber, Lime shareable scooters, and messaging system Slack.

Capital G: A late-stage growth venture capital fund financed by Alphabet and based in San Francisco. Founded in 2013, it focuses on larger, growth stage technology companies. It invests for profit rather than strategically for Google. A few notable investments include ride sharing company Lyft, financial payment company Stripe and hospitality company Airbnb.

Jigsaw: (formerly Google Ideas) is a technology incubator originally created by Google in 2010. The incubator aims to solve the toughest global security challenges facing the world today including thwarting online censorship, mitigating digital attacks, and protecting people from online harassment.

Verily: A research organization devoted to the study of life sciences. The organization was formerly a division of Google X, until 2015 when Sergey Brin announced that the organization would become an independent subsidiary of Alphabet Inc.

DeepMind: Founded in 2010 in England and acquired by Google in 2014. DeepMind has created a neural network that learns how to play video games in a fashion similar to that of humans. The mission statement is to research artificial intelligence and its applications for positive impact.

Nest: Palo Alto-based manufacturer of smart home products including thermostats, smoke detectors, and security systems. Next was acquired by Google in 2014.

Waymo: Formerly the Google self-driving car project—Waymo stands for a new way forward in mobility. Most agree Waymo is the most advanced in the field of autonomous vehicles.

Calico: The California Life Company, Calico for short, is Alphabet's anti-aging research laboratory. Calico studies what makes people age and develop tools to help people live longer and healthier lives. It is run by doctors, drug experts, molecular biologists, and geneticists.

Google Fiber: Provides fiber-to-the-premises service in a number of U.S. cities including Kansas City, Austin, Provo, Atlanta, Charlotte, Nashville, Salt Lake City, and San Antonio.

Sidewalk Labs: is Alphabet's urban innovation organization which aims to apply technology to help solve urban infrastructure problems. The goal is to tackle issues such as cost of living, efficient transportation and energy usage.

Chronicle: A cybersecurity company which was originally developed by Google X and became its own company under Alphabet in 2018. Chronicle creates tools for businesses to prevent cybercrime on their platforms.

eBAY

eBay founder Pierre Omidyar and former CEO Meg Whitman.
Omidyar served as chairman from 1998 to 2015. eBay

eBay's campus in San Jose has named their buildings after the categories of the items they sell.

eBay was founded in 1995 by Pierre Omidyar under the name AuctionWeb. Omidyar tested his new auction website by posting a broken laser pointer for sale. To his surprise, a collector bought it for $14.83. So began a radical transformation in commerce. In less than a year and with only a few employees, eBay was seeing $10,000 per month in revenue and had 41,000 registered users. eBay's growth continued exponentially from there, and by 1997 they were hosting 200,000 auctions per month: ten times the volume of just a year before. In 1998 eBay went public. By 1999 there were 10 million registered users.

The most expensive item ever sold on eBay? A 405-foot yacht complete with a movie theatre on board. Highest bid: $168 million.

Devin Wenig is eBay's President & CEO. He is the third CEO in the company's 20+ year history.

In 1997, eBay sold its one millionth item: A Big Bird toy.

In 2012 an eBay auction for lunch with Warren Buffett raised over $3 million for charity.

In 2005, the original Hollywood Sign was sold on eBay for $450,400.

In 2004, a 10-year-old grilled cheese sandwich said to bear the image of the Virgin Mary sold for $28,000.

Other bizarre purchases include:

- In 2001, Ellen DeGeneres auctioned off a lock of Justin Bieber's hair for over $40,000. The money was donated to charity.
- In 2009, A cemetery crypt located directly above Marilyn Monroe was auctioned off for $4.6 million. The crypt next to Monroe was purchased by Hugh Hefner in 1992 for $75,000.
- In 2009, a 22-year-old woman from San Diego attempted to sell her virginity on eBay although the auction was suspended by the company.
- In 2006, William Shatner sold his kidney stone for $75,000.

Reid Hoffman

Reid Hoffman in 2011. Wikimedia

IPO of LinkedIn in 2011. CEO Jeff Weiner is standing at the podium, Reid Hoffman is to the left. LinkedIn

Reid Hoffman was born in Palo Alto and grew up in Berkeley. He graduated from Stanford University in 1990 with a B.S. in Symbolic Systems and Cognitive Science. During his sophomore year, he met and befriended classmate Peter Thiel. Hoffman went on to receive his Masters from Oxford University. He joined Apple Computer in 1994 and helped develop eWorld. eWorld was Apple's response to AOL's e-mail and chat-room service. Hoffman built on what he had learned and in 1997 he co-founded his first company, Socialnet.com. The vision was to connect people with similar interests—it was a social network years before that became a buzzword. In 1998, Thiel tapped Hoffman to join the board of directors at his fledgling company—PayPal. Two years later Hoffman joined PayPal full-time as the company's COO. When Paypal was acquired in 2002, Hoffman again tried his hand at social networks and founded LinkedIn. He served as CEO for the first four years and then served as Chairman of the Board until its sale in 2016 at which point he joined Microsoft's board. In 2010 Hoffman joined Greylock Partners to further his already prolific angel investing. Hoffman and Thiel were the original angel investors in Facebook. Hoffman has also invested in Airbnb, Nauto, Digg, Flickr, IronPort, Nanosolar, Shopkick, TrialPay, among many others. In an effort to spread his vast knowledge on what it takes to create a wildly successful company, Hoffman now hosts a podcast called Masters of Scale. Each episode he tests his theories on what is required for success on legendary Fortune 500 company leaders.

LinkedIn

The LinkedIn team posing in front of their Mountain View campus at the 5 million user milestone. Founder Reid Hoffman can be seen in the back right. LinkedIn

The LinkedIn team posing in front of their Mountain View campus at the 150 million user milestone. LinkedIn

LinkedIn was founded in 2002 by Reid Hoffman and a team of engineers assembled from PayPal and Hoffman's other start-up, Socialnet.com. The series A round of investment came from Sequoia Capital in 2003. The first few years of operation were marked by exponential growth. Within the first month of operation, the site had 5,000 members; within a year it was up to 1 million, and by year four it had 10 million users. When LinkedIn went public in 2011, its share price doubled in the first day of trading, jumping from $45 to $94. In 2016, Microsoft acquired LinkedIn for $26 billion—the equivalent of $196 a share. Microsoft has allowed LinkedIn to retain its distinct brand, culture and independence with Jeff Weiner continuing as CEO. As of 2018, LinkedIn had 590 million registered members. As opposed to most other social networks, LinkedIn's business model doesn't depend strictly on advertising revenue. In 2016, 65% of its revenue came from recruitment services, 17% from premium subscriptions and only 18% from advertising. This is in sharp contrast to Facebook which reports 98% of its income is from advertising.

MARK ZUCKERBERG

Mark Zuckerberg circa 2005. Wikimedia

Mark Zuckerberg in his Palo Alto office. Wikimedia

Mark Zuckerberg was born in New York in 1984. As an undergraduate at Harvard University, Zuckerberg came up with Facemash, an online game for Harvard students that allowed participants to compare two student pictures side-by-side and choose who was "hot" and who was "not." The website gained overnight popularity with students but was soon shut down by Harvard executives. Zuckerberg faced charges of violating copyrights, breach of security, and violating individual privacy, for which he faced expulsion from the University. Charges were eventually dropped. In 2004, Zuckerberg launched thefacebook.com, a social network initially only available to Harvard students, but later expanded to all Ivy League schools including Stanford University. Shortly after the launch of thefacebook.com, three Harvard students (Cameron & Tyler Winklevoss and Divya Narendra) accused Zuckerberg of intentionally misleading them into believing that he would help them build a social network called HarvardConnection.com. They claimed that instead, he used their idea to build a competing product. They sued and in 2008 were awarded $300 million in Facebook stock. Much of the drama was detailed in the 2010 film, *The Social Network*.

Facebook

According to the New York Post, Facebook owns 4 of the 5 most downloaded apps. Facebook is the number one downloaded app, followed by the Facebook messenger (second), YouTube (third), Instagram (fourth) and WhatsApp (fifth).

In the summer of 2004, Facebook incorporated and relocated to Palo Alto. Zuckerberg tapped Sean Parker, the man behind Napster, to become the company's first president and help raise the necessary capital. Sean Parker initially approached Reid Hoffman, the founder of LinkedIn. Hoffman directed Parker to Peter Thiel who provided the first angel investment of $500,000. In 2006, Facebook granted access to the site to anyone over the age of 13. By 2011, Facebook was second only to Google in site visits. In early 2012, Facebook acquired photo-sharing app Instagram for $1 billion—a bold move considering the 13 person start-up had absolutely no revenue. Later that year, Facebook went public. Although their public offering was plagued with technical problems and a lack of public support, the stock recovered shortly thereafter and experienced years of appreciation. In 2014, Facebook bought messenger app WhatsApp for the hefty price-tag of $19 billion and later that year acquired virtual reality company Oculus virtual reality for $2 billion. As of 2018, 68% of Americans have a Facebook account and 43% get their news from the social media site. The company has come under fire for not taking their role as a media company seriously and allowing the dissemination of fake news. Zuckerberg asserted that they are a technology company, saying "We are the platform, not the publisher." Some argue that Facebook's hands-off approach to policing inaccurate or incendiary content allowed foreign agents to impact the results of the 2016 US presidental elections. Moving forward, Facebook and the governmental agencies tasked with overseeing this threat will need to figure out how to moderate inappropriate content while not infringing on citizens' right to free speech.

PayPal

In 1998, Peter Thiel and Max Levchin met for breakfast at Hobee's restaurant in Palo Alto to discuss Levchin's idea—to beam money between PalmPilots. Shortly thereafter, a company was born. Its milestone product was a worldwide online payment system called PayPal. In 2000, Thiel and Levchin's company merged with another online banking company founded by Elon Musk. Shortly after the merger, Thiel resigned.

Peter Thiel and Max Levchin in 2002, celebrating PayPal's first day of being a public company. Max Levchin

Musk became CEO of the combined company and decided it was time for a technological overhaul. Specifically, he wanted to do away with Unix and put everything on a Microsoft platform. A holy war ensued. Musk lost. The board fired him and brought back Thiel while Musk was on a flight to Australia for his first vacation in years. After a corporate restructuring in 2001, the company formally adopted the name PayPal and made the decision to use eBay as a key platform to grow its user base.

Peter Thiel and Elon Musk circa 2000. Associated Press

CEO Dan Schulman (front & center) and the PayPal team in 2015 celebrating the company's return to being an independent, publicly traded company. NASDAQ.

PayPal went public in February 2002 only to be acquired by eBay later that year for $1.5 billion—77% above the IPO price. In the early years, PayPal had a large list of competitors which included: eBay's subsidiary Billpoint, Yahoo!'s PayDirect, Google Checkout, Western Union's BidPay service, Citibank's c2it, all of which are now defunct. By 2010 PayPal had over 100 million active user accounts in 190 markets through 25 currencies. In 2013 PayPal acquired Venmo—a mobile payment app popular with millennials—as part of its $800 million acquisition of Baintree. In 2014 it was announced that Dan Schulman would become CEO of PayPal. Schulman previously served as the CEO of Priceline and Virgin Mobile. He continues to serve as the Chairman of Symantec. Shortly thereafter PayPal was spun out from eBay and again operates as a separate legal entity, going public again in 2015. As of 2018 PayPal has 244 million active accounts worldwide, is available in 100 currencies and processed $139 billion in total payments in the second quarter of 2018.

PayPal Mafia

Veterans of the early days of PayPal pose for a group shot at Tosca in San Francisco in 2007. This photograph is largely the reason the term PayPal Mafia became well known. Back row from left: Jawed Karim, Jeremy Stoppelman, Andrew McCormack, Premal Shah. 2nd row from left: Luke Nosek, Kenny Howery, David Sacks, Peter Thiel, Keith Rabois, Reid Hoffman, Max Levchin, Roelof Botha, Russel Simmons. Absent: Elon Musk, Chad Hurley and Steven Chen. Robyn Twomey for Fortune Magazine.

The PayPal Mafia is a term given to the early employees of PayPal who have gone on to found and fund many additional successful technology companies. Former PayPal CEO Peter Thiel estimates the PayPal Mafia to be around 220 people who have advised and invested in over 400 tech companies. These serial entrepreneurs and investors are the twenty-first-century version of the Traitorous Eight. They are classic characters of Silicon Valley and are the reason people flock to the Bay Area from all over the world. The PayPal Mafia includes three self-made billionaires, a host of millionaires and a legacy of generation-defining companies including:

TESLA

Elon Musk led the Series A venture capital round for the company in 2004 and at that point became Chairman of the Board. Tesla has a $50 billion market capitalization as of 2018.

LinkedIn

Founded by Reid Hoffman and other PayPal alumni in 2002. LinkedIn was purchased by Microsoft in 2016 for $26 billion.

Palantir

Co-founded by a group of Stanford Alumni including Peter Thiel and Joe Lonsdale. Palantir has an estimated value between $6–$20 billion.

SPACEX

Founded by Elon Musk in 2002. SpaceX has an estimated value of $27 billion.

yelp

Founded in 2004 by Jeremy Stoppelman. Yelp has a $4.26 billion market capitalization as of 2018.

Affirm

Founded by PayPal co-founder Max Levchin in 2012. Affirm has an estimated value of $2 Billion as of 2017.

YouTube

Founded in 2005 by Chad Hurley, Steve Chen and Jawed Karim while still working at PayPal. YouTube was acquired by Google in 2006 for $1.65 billion.

yammer

Founded by David Sacks and Adam Pisoni in 2008. Yammer was acquired by Microsoft in 2013 for $1.2 billion.

PALANTIR

On the left: Alex Karp, CEO of Palantir and a self-proclaimed neo-Marxist. On the right: Peter Thiel, founder of Palantir and a self-proclaimed Libertarian. Alex & Co.

Heard of Palantir? Maybe not, but they've heard of you. The Palo Alto based company has created a software program that compiles data across many platforms including cell phone records, social media, financial documents, airline reservations, GPS coordinates, browser history and toll both cameras. Their algorithm then generates a visual representation of how everything is connected in easy-to-read color-coded diagrams that look like spider webs. Palantir was founded in 2004 by former PayPal founder Peter Thiel. Thiel tapped his Stanford law school classmate Alex Karp to be CEO and hired various other Stanford and PayPal alumni as initial employees. Palantir got its start building upon the software originally developed to predict potential fraudulent users of PayPal. Similar to the early semiconductor industry, Palantir's first investor and client was the US Government. The venture capital arm of the CIA, In-Q-Tel, provided $2 million in start-up money. Palantir's technology was first deployed in the fight against terrorism in Afghanistan and Iraq. It was used to track insurgents for assassination including Osama bin Laden. Palantir's success on the battlefield led to more government contracts, this time on the civilian side. The Department of Health and Human Services picked it up to detect Medicare fraud; the Department of Homeland Security now uses it to screen air travelers for terrorists and illegal immigrants; local law enforcement use it to track gang activity. Soon, corporate America also wanted in—Palantir's client list includes financial institutions such as JP Morgan Chase, airplane maker Airbus, and pharmaceutical giant Merck.

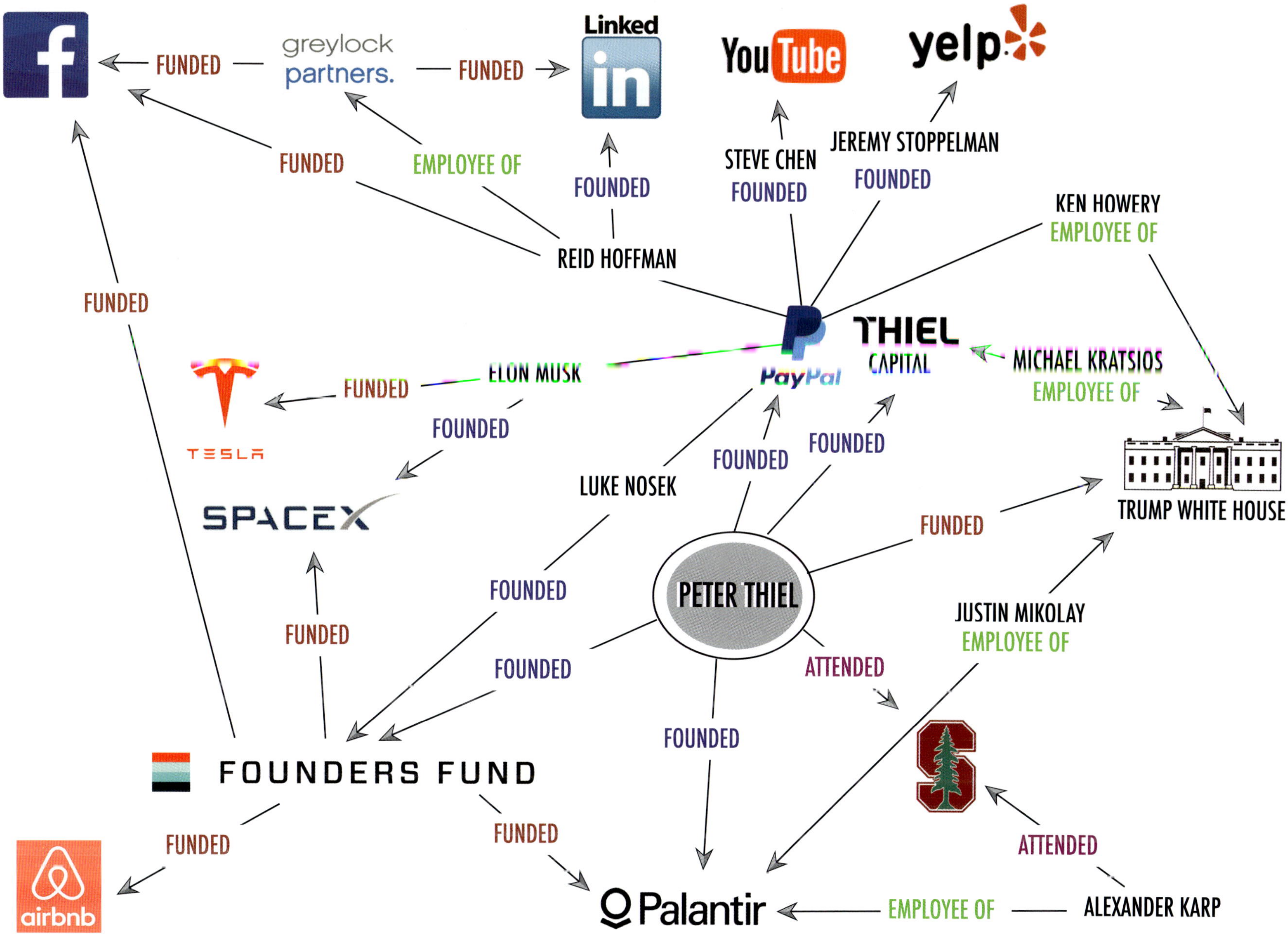

A graphic representation of how the life of Peter Thiel, founder of Palantir, might appear in the company's model. The idea for this graphic is based on one originally published by Bloomberg News.

The company has faced criticism in recent years, some accusing the company of being a spy and an agent for big brother. Opponents say the software once designed for the global War on Terror has now been "weaponized against ordinary Americans." Palantir counters that it is up to the user on how the software is deployed telling Bloomberg news: *We are aware that powerful technology can be abused and we spend a lot of time and energy making sure our products are used for the forces of good.* According to Crunchbase, Palantir has raised close to $2 billion in funding as of late 2018. However, many analysts are unsure how to value the company due to the highly secretive nature of their business—as of early 2018 Palantir had yet to report a profit.

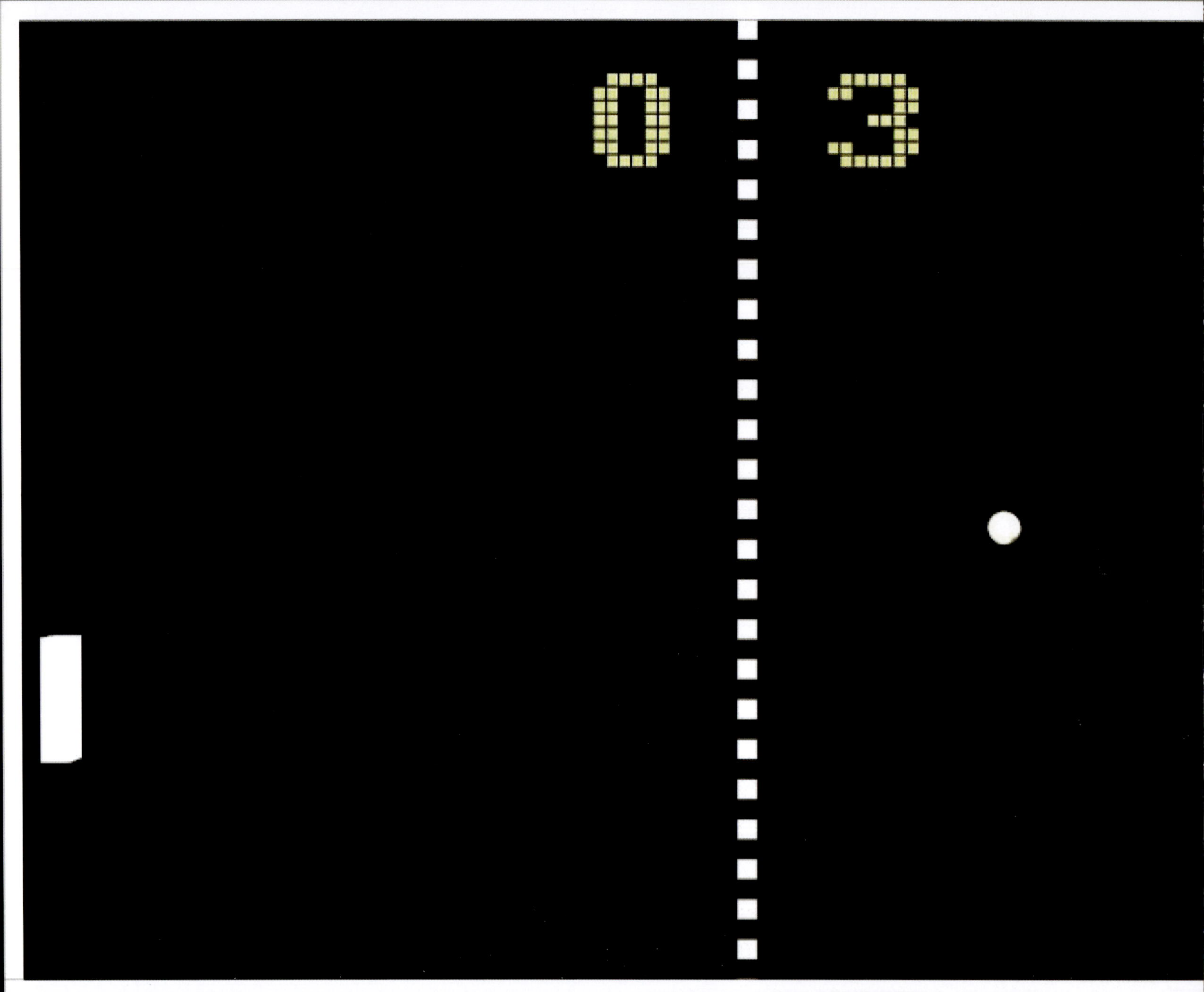

Created in 1972, Pong was the first computer game produced by Atari.

Early mainframe computers were once only available to those with deep pockets—namely the government, large corporations, and research universities. Although as processing power improved, computers were no longer only reserved for mission-critical operations. Visionaries imagined how the new technology being created in Silicon Valley could be applied to other industries. It wasn't long before computers were programmed strictly for entertainment. The first computer video game was named Spacewar. It was originally developed in the late 1960s at MIT by Steve Russell. The game made its way to the Valley when Russell joined the Stanford University Artificial Intelligence Lab. Stanford University then sponsored the first computer game tournament using the Spacewar game in 1972. Rolling Stone magazine hailed the event as a pivotal moment in popular culture. Six weeks later, Atari released the world's first commercial video game, Pong. Since then, Silicon Valley has spawned many companies whose main goal has been to harness technology to change the face of entertainment.

Fourteen: Entertainment

ATARI

Founders Ted Dabney and Nolan Bushnell with Fred Marincic and Al Alcorn. While Dabney and Bushnell founded Atari in 1972, both would be gone from the company by 1980. Dabney left in 1973, believing the market was unstable; Bushnell left in 1979, after selling the company to Warner Communications.

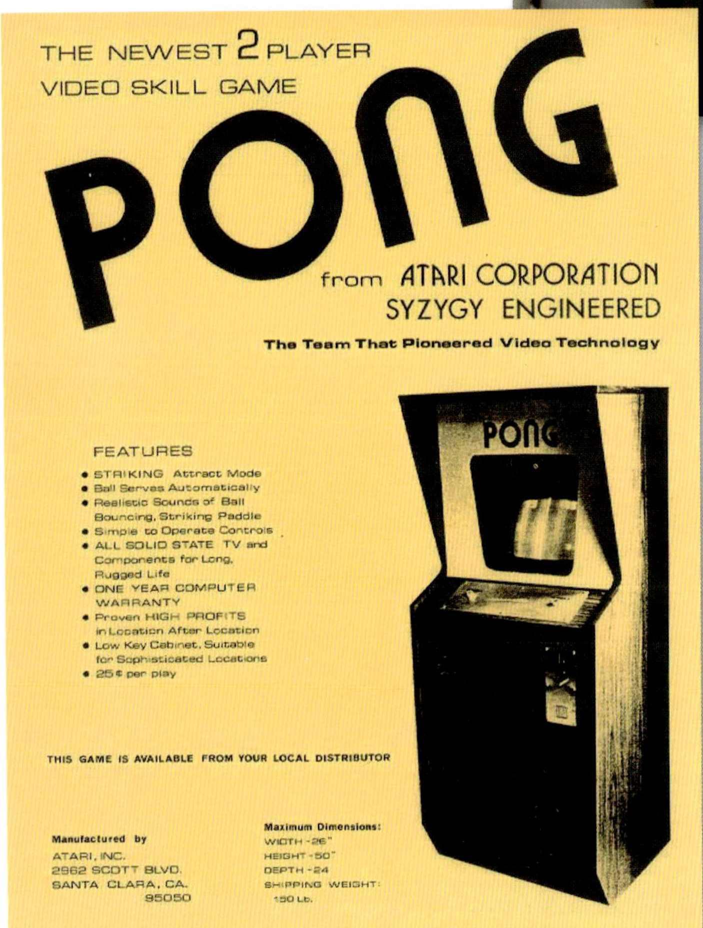

In 1971 Nolan Bushnell and Ted Dabney designed and built the first arcade video game: Computer Space. In 1972 the pair founded Atari and soon hired Al Alcorn as their first design engineer. Bushnell decided to have Alcorn produce a game to test his abilities. From this training exercise came the arcade game Pong, an animated table tennis game. Atari placed the prototype in Andy Capp's Tavern in Sunnyvale to gage the public response. Legend has it that within the first week the bar owner called to complain that the machine was broken. When Alcorn investigated, he discovered the real problem was that the game was too popular—the coin box was full and preventing further use. Needless to say, Pong was an overnight success and marked the start of the video game industry.

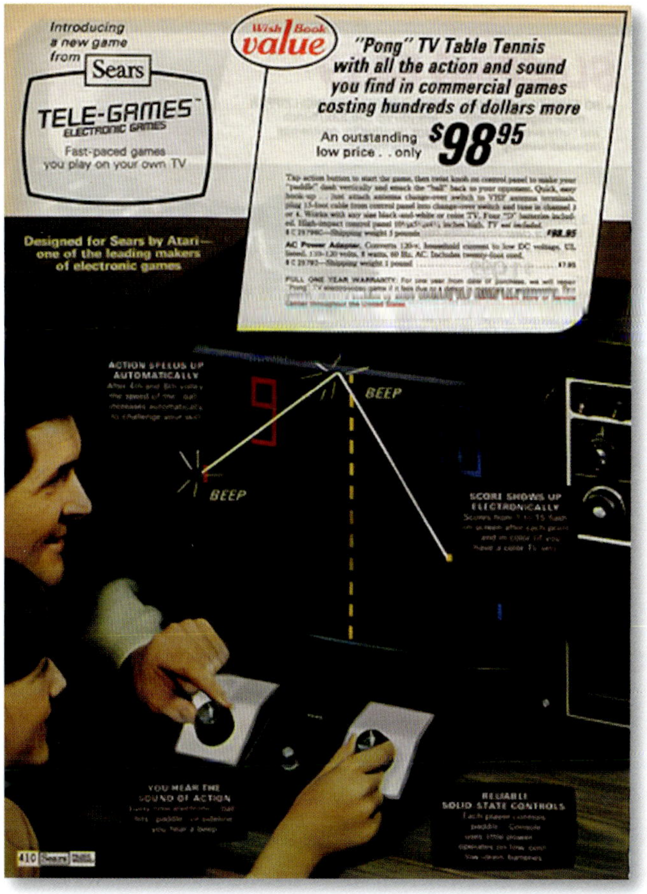

Pong advertisement in the Sears catalog.

Sears advertisement promoting a later version of the Atari video game console.

Nolan Bushnell is a larger than life character, and served as the company's spokesman. After the initial success of Pong, he went out in search of new capital to help propel the growth of his company. The problem he kept running into was that most investors saw the arcade game business as too capital intensive because large inventories were needed to support the business model. Then, Nolan met Don Valentine, who tells the story that when he went to check out Atari's assembly line in Los Gatos, he was nearly knocked over by the smell of marijuana. Undeterred, Valentine saw the real potential of the company to be in the recently developed, but not yet produced, home video console. He saw this console as having the ability to open up a new market segment and, assuming you could sell large quantities to retailers, could eliminate the need to house large inventories. He provided the funding Atari needed. However, once the home console was produced, Bushnell found that retailers, including toy stores and electronics stores, didn't see the value in the product. This problem almost took the company down, until Valentine reached out to a friend who was a large investor in Sears. Thankfully, this call helped facilitate the exclusive distribution agreement between Atari and Sears, which was just in time for the 1975 Christmas season. To Bushnell and Valentine's delight, the home console was met with overwhelming demand from consumers.

Electronic Arts

Trip Hawkins, the founder of Electronic Arts circa 1985.

Trip Hawkins, the founder of Electric Arts, says that while the company was officially founded in 1982, the true genesis dates back to 1970 when he was 16 years old. Around this time, he noticed how some of his friends were drawn to nerdy games involving math while others preferred just watching television. This got him thinking—was there a way to combine these two activities? Hawkins left Apple in 1982 and used his own money to provide the capital for EA's first year of operation. Don Valentine of Sequoia offered him the use of some office space. They set up card tables at the end of the hallway to accommodate the team. Hawkins says that while working at Apple around geniuses like Bill Atkinson, Larry Tesler, and Bill Budge, he realized that software developers are artists. It gave him the idea to make EA the first software company to use words like artist, producer, and director. Electronic Arts became famous for their game packaging, styling them after record albums. Hawkins recruited his original employees from Apple, Atari, Xerox PARC, and VisiCorp. He got Steve Wozniak to sit on the board of directors. Some of the early games included Pinball Construction Set, M.U.L.E. and Archon. Hawkins served as CEO and Chairman of the Board from 1982 through 1994.

Pinball Construction Set was a video game developed by Bill Budge and published by Electronic Arts in 1983. In this game, users would construct a virtual pinball machine by placing bumpers, flippers and other pinball paraphernalia onto a table.

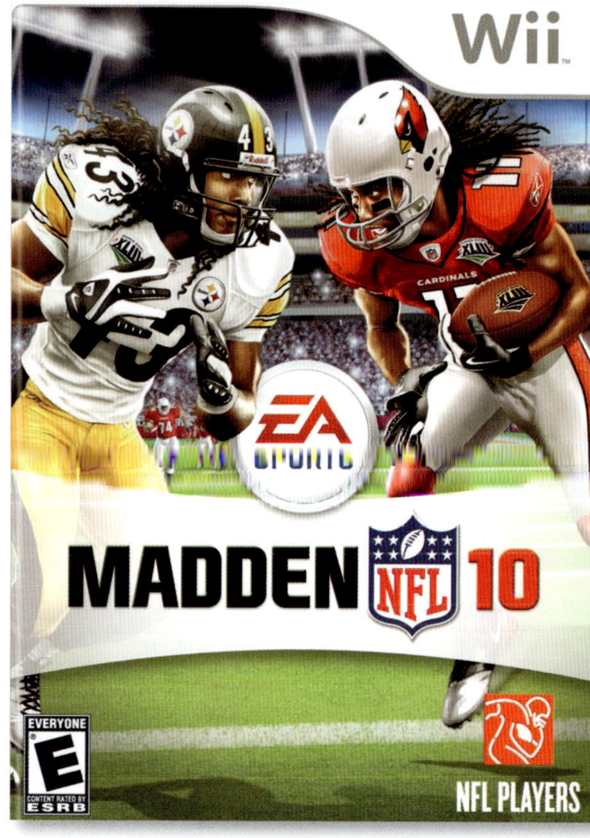

First released in 1988, Madden Football is a video game using the likeness of real NFL players and has voice commentary so players hear the game being called as if it were a real game on TV.

In 1984, Larry Probst joined the company as the VP of Sales and was later promoted to CEO by Trip Hawkins. He still serves as the Chairman of the Board. The company continues to develop, produce and/or license new games. One popular innovation has been to partner with major league sports organizations like the NFL, NHL, NA and PGA to bring current day sports figures into the video game realm. CEO Andrew Wilson joined EA in 2000 and became CEO in 2013. As of September 2017, Electronic Arts is the second-largest gaming company in the Americas and Europe by both revenue and market capitalization.

Electronic Arts purchased land in Redwood Shores to construct a new headquarters completed in 1998 and is still occupied today.

TIVO

The TiVo team in the early 2000s. TiVo.

Initially going by the name Teleworld, TiVo was founded in 1997 by Mike Ramsay and Jim Barton. The duo were inspired after working together on a joint venture between Time-Warner and Silicon Graphics to create the first interactive television system. The first prototype of TiVo's digital video recorder which digitized and compressed analog video from any source (antenna, cable or satellite) was unveiled at the Consumer Electronics Show in January 1999. At the conference Ramsay announced that the company would be ready to ship the first-ever DVR by March 31, 1999. Despite the tight deadlines and other challenges, TiVo engineers managed to be on schedule. In honor of the first successful launch, the company continues to celebrate March 31st as a company-wide holiday. The DVR set top boxes were a tremendous hit with the public and TiVo went public only 6 months later in September 1999. It took until 2005 for the company to reach profitability.

Jim Barton and Mike Ramsay shipping out the original TiVo DVR in 1999. TiVo.

The wall of innovation in the new TiVo headquarters building in Alviso. TiVo.

In June 2005, veteran media executive Tom Rogers took the reins from TiVo co-founder Mike Ramsay. Rogers served as president and CEO until 2016 when Rovi acquired TiVo Inc. for $1.1 billion. The combined company has continued to operate under the TiVo brand and holds over 6,000 patents.

NETFLIX

Founder and CEO of Netflix, Reed Hastings in the Silicon Valley DVD distribution center, circa 2002. Associated Press.

Netflix was founded in 1997 by Reed Hastings and Marc Randolph. Randolf worked for Hastings at his previous start-up company. The pair shared a commute from Santa Cruz to Sunnyvale and on the drive would discuss new business ideas. Randolph was intrigued with Amazon, then a fledgling online book retailer. He wanted to explore pursuing a similar business model and was initially focused on selling VHS tapes over the Internet. They determined, however, VHS tapes were too expensive to stock and too delicate to ship. A lucky coincidence, the DVD was introduced right around the same time. The DVD proved easy to stock and sturdy enough to mail. Lore has it that Hastings attributes the start of Netflix to the frustration over a $40 Blockbuster late fee. With a modest seed investment ($2.5 million of which came from Hastings' windfall from the sale of his previous company) and 30 employees, the pair set off to take on the $16 billion home video rental industry. Randolph served as the CEO for the first year of operation. Hastings took over as CEO in 1999, a post he continues to hold today.

Netflix headquarters building in Los Gatos. Netflix

Netflix headquarters building in Los Gatos. Netflix

When Netflix originally launched as the first online DVD rental store, they offered every DVD available at the time—925 titles. Their video rental rates and rental period were similar to brick and mortar video stores. Two years later Netflix dropped the single-rental model and focused exclusively on a monthly subscription model. This gave users unlimited rentals without due dates or shipping fees. By early 2001, Netflix had acquired 300,000 subscribers but was still operating at a loss. Fortuitously, the cost of DVD players dropped sharply that year becoming the go-to Christmas gift and giving a large boost to the popularity of the Netflix subscription model. Netflix went public in 2002 and posted their first profit in 2003. While working in product development at Netflix, Randolph and another founding team member, Mitch Lowe (who currently serves as the CEO of MoviePass) developed an idea for a movie rental kiosk called Netflix Express. Netflix later rejected this line of business and Lowe acquired the rights to the kiosk concept and turned it into Redbox. Randolph retired from Netflix in 2004. By 2005, Netflix was shipping out 1 million DVDs every day and offered 35,000 film titles. As Internet speeds increased, Netflix began streaming content over the Internet. By 2010 Netflix's streaming business was the largest source of Internet traffic in North America in the evening. In 2013, Netflix produced its first series, House of Cards. With this move into original content, Netflix chief content officer Ted Sarandos told *GQ* the plan was "to become HBO faster than HBO can become us." Sure enough, barely three years after it started producing its own content, Netflix had more original series in production than HBO. In May of 2018, Netflix stock hit an all-time high giving them a market capitalization of $152.7 billion; briefly surpassing Disney ($152.3 billion) and Comcast ($142.6 billion).

ROKU

Roku offers the ability to stream content via a set-top box, Roku smart TV or on an independent device such as an iPhone. ROKU.

Roku—meaning six in Japanese—was the sixth company founded by Anthony Wood. To back up...Wood joined Netflix in 2007. Around that time, Netflix was debating whether to offer a device to exclusively stream Netflix content. It was finally decided by Netflix CEO Reed Hastings not to pursue this line of business. Netflix wanted to encourage other companies, like the manufacturers of game systems and DVD players, to include access to Netflix content on their devices. In the end, Netflix spun off their set-top box division into its own company that would be called Roku with Netflix as an early investor. Netflix then pursued making its content available on Roku and all other devices. Roku was then free to offer video programs not just from Netflix but also from content providers like Hulu, Amazon and YouTube on this new streaming device.

Roku founder Anthony Wood (center) rings the Nasdaq stock market opening bell on September 28, 2017. Roku stock jumped 67% on its first day of trading and was considered Silicon Valley's top performing IPO of 2017. NASDAQ.

In 2014, Roku announced its first branded Smart TV. This allowed third-party television manufacturers to market a TV with the Roku user interface acting as the brain. Roku has faced competition from Amazon, Apple and Google. However, according to a survey in 2017 Roku continues to be the number one TV streaming platform in the United States as measured by total hours streamed. Roku went public in 2017, jumping 67% on its first day as a public company. It was considered Silicon Valley's top performing IPO of 2017. Originally headquartered in Saratoga and then next door to Netflix in Los Gatos, in 2018 Roku signed a lease for nearly half a million square feet in a new high-rise office complex next to the San Jose Airport on Coleman Avenue.

YouTube

YouTube founders from left to right: Chad Hurley, Steven Chen and Jawed Karim.

YouTube was founded by Chad Hurley, Steve Chen, and Jawed Karim—all early employees of PayPal. There are a few legends on where the idea for the video-sharing company originated. One is that the trio experienced difficulty sharing videos from a dinner party. Another is that inspiration came from the inability to find a video online of Janet Jackson's exposed breast during her 2004 Super Bowl performance. Yet another is that the original intention was to create a video version of an online dating service, influenced by the website Hot or Not. The domain name www.youtube.com was activated in 2005 and the first video, titled *Me at the zoo*, shows co-founder Jawed Karim at the San Diego Zoo and can still be viewed on the site. YouTube's initial headquarters was above a pizzeria and Japanese restaurant in San Mateo. By the summer of 2006 YouTube was already averaging 100 million video views per day—mostly by 12 to 17 year-olds. Later that year, Time Magazine featured a YouTube screen with the word "YOU" as its annual 'Person of the Year' highlighting the rise of user-created media.

CEO of YouTube, Susan Wojcicki is seen here in a YouTube clip on how to properly pronounce her name. It's wah-JIS-kee.

Google was founded in Susan Wojcicki's garage in 1998. Wojcicki was Google's first marketing manager starting in 1999. Wojcicki was later put in charge of Google's original video service. She is credited with convincing Google to acquire YouTube in 2006 for $1.65 billion. In 2014 Wojcicki became the CEO of YouTube, a role she continues to hold. Today, there are 1,000,000,000 mobile video views per day. South Korean pop star video Gangnam Style became the first video to clock more than one billion views on YouTube and held the title of most watched video for 4.5 years until losing the top spot in 2017. Cumulatively, viewers watch more YouTube than Netflix and Facebook video combined.

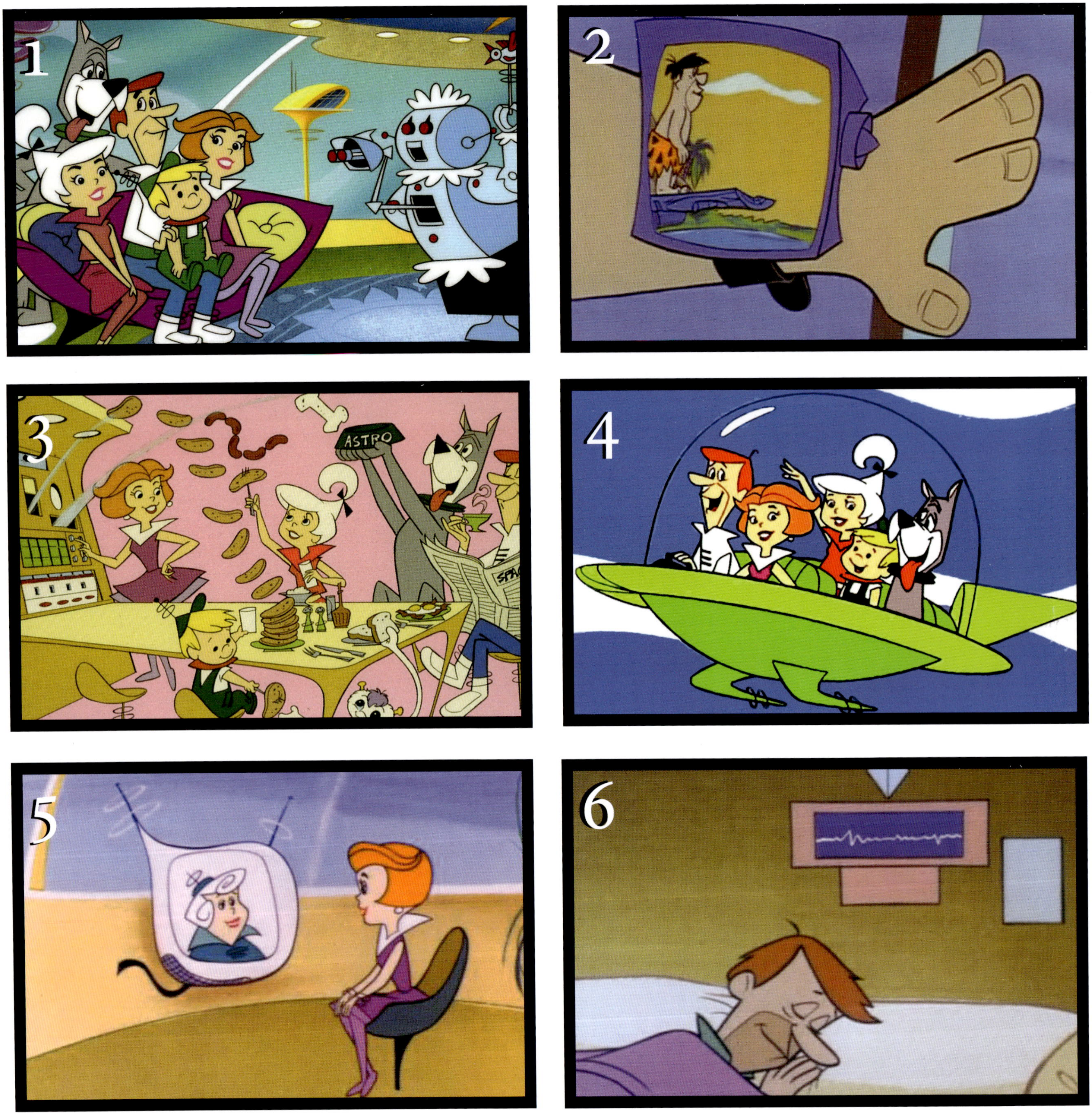
1
2
3
ASTRO
4
5
6

As computing power continues to advance and Internet bandwidth expands, the "inventions of the future" are become a reality. Artificial Intelligence, robots, self-driving cars, and Internet-connected devices are transforming the way we interact with the world. Many inventions and technology being developed in Silicon Valley today were strictly in the realm of science fiction fifty years ago. The Jetsons originally premiered in 1962 and features a family living in the year 2062. The technology coming out of Silicon Valley has made many of the Jetsons' futuristic dreams a reality. Highlights include.

1—ROBOTS: A robot with all the capabilities of Rosie the Robot might still be a few years out but we are getting close. Today we have robots which can perform some of Rosie's household chores. Roomba, the automated vacuum cleaner, is a great example.

2—APPLE WATCH: In one episode, an elementary school student slacks off watching The Flintstones on his wrist TV. His device is strikingly similar to today's Apple Watch.

3—CONNECTED DEVICES: Today's smart kitchen devices connected to a network could hypothetically recommend and display a recipe for lasagna because the ricotta cheese is close to its expiration date. The fridge could then alert the oven to preheat to the proper temperature and let the dishwasher know the dishes will need a heavy cycle.

4—UNMANNED DRONES & FLYING CARS: In the opening credits, George Jetson transports his family in a flying car and drops each member off via individual pods which take them to their final destination. Today we would call those pods unmanned drones. There are many commercial drones under development promising to deliver everything from blood samples to your latest online purchase. Sebastian Thrun is working on a prototype of a personal aircraft.

5—FACETIME: The show features various examples of what a videophone could look like. Facetime now allows anyone with an Apple device and an Internet connection the ability to hold a video conference.

6—SIRI: The Jetsons featured a talking alarm clock which was voice operated. Today we have a host of voice-activated apps which provide users a virtual assistant which will perform a wide range of tasks including setting alerts and reminders.

This chapter explores the cutting-edge technology and wide reaching applications of Artificial Intelligence, Internet of Things (IoT), Global Positioning System (GPS), robots, and self-driving cars.

Fifteen: Into The Future

IoT — Internet of Things

Les Earnest is a computer scientist and visionary who worked at SAIL (Stanford Artificial Intelligence Laboratory) between 1965 and 1988. Earnest was involved with inventions perhaps so far ahead of their time that they were largely unappreciated until later. One such invention in 1971 was the first computer operated vending machine. The vending machine was hooked up to a teletype. A user would type in what he or she wanted, the computer would open the door for the customer to remove the item and then charge the purchase to the user's account. At one time the machine also sold beer—if a customer was underage the display would read, 'Sorry, kid.' The vending machine was nicknamed the Prancing Pony in reference to the Lord of the Rings trilogy. In the early eighties, students at Carnegie Mellon became inspired to create their own connected vending machine. They duplicated the design of the Prancing Pony and then took it a step further. They connected the coke machine to the ARPANET—a precursor to today's Internet—so they could monitor the contents remotely. Many people point to this as the first non-traditional, everyday item to be connected to a network, kicking off the Internet of Things revolution.

At its most basic level, IoT is the concept of connecting any device that has an on/off switch to the Internet and to each other. This includes everything from cellphones, coffee makers, washing machines, and headphones, to wearable devices, toilets, traffic lights, parking spots and even ingestible sensors. These connected devices throw off a massive amount of data. In 2012 IBM calculated that every person on the planet creates 500 megabytes of data every day. If we tried to back up just one day's worth of data—printed it out on double-sided 8x11 paper, size 12 font—the stack of paper would reach from the earth to the sun and back four times over. But that is just the beginning. Analyst firm Gartner estimates that in 2017 there were approximately 8 billion connected devices. They estimate this will be well over 26 billion by 2020. At that point, there will be four times more connected devices on the planet than people.

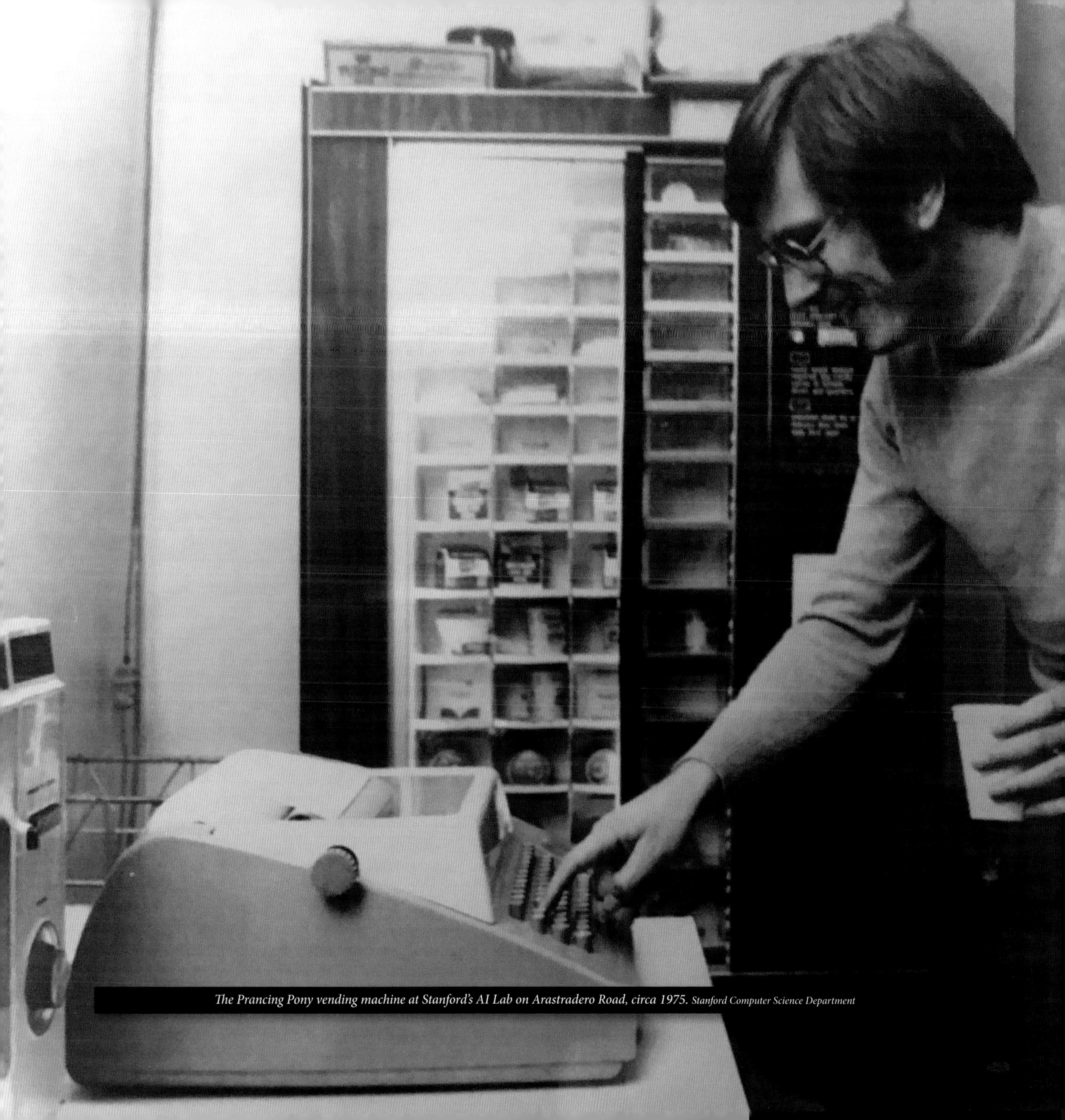

The Prancing Pony vending machine at Stanford's AI Lab on Arastradero Road, circa 1975. Stanford Computer Science Department

GPS

Artist rendering of GPS satellite orbiting Earth. Nasa for Wikimedia

GPS, or Global Positioning System, is a worldwide positioning, navigation and timing service and is a major component of IoT technology. With GPS technology now built into most smartphones and navigation systems for new cars, it's hard to remember this technology is still relatively new. As with most revolutionary technology, the satellite-based positioning system was developed by U.S. Department of Defense. In 1978, the first NAVSTAR satellite was launched into space. Today, the GPS system is formed from a constellation of at least 24 satellites and their ground stations, which are used for both military and civilian uses. The satellites orbit 12,550 miles above the earth and circle the globe twice a day. On a basic level, GPS works by triangulating your position on the ground—it calculates how far away you are from each satellite and how far apart each satellite is from one another. Overtime, GPS receivers have been miniaturized down to a single integrated circuit, which is why the technology is part of virtually every facet of modern life—from wristwatches to bulldozers, shipping containers, and ATM's.

TRIMBLE

Yak-mounted GPS receiver surveying Mt. Everest in the early 1990s. Trimble.

In the early days, a GPS receiver weighed around 50 lbs and cost approximately $150,000. Trimble.

World's first in-car navigation system with CD maps in 1990 used Trimble GPS. Trimble

3D survey of the Golden Gate Bridge after Loma Prieta Earthquake. Trimble

Trimble Navigation, now known simply as Trimble, was founded by Charlie Trimble and two other engineers from Hewlett Packard in 1978—the same year the first GPS satellite was launched into space. Charlie recognized the potential of GPS technology and foresaw a market for commercial and consumer applications, not just military ones. Feeling inspired, the young company purchased Hewlett Packard's immature GPS technology and introduced its first GPS product in 1984. In 1990, Trimble became the first publicly traded GPS company. Steve Berglund joined Trimble in 1999 as President and CEO—a role he continues to hold today. Trimble's GPS technology has transformed numerous industries including agriculture, construction, transportation and logistics. Today, Trimble offers over 500 products with core technologies in positioning, modeling, connectivity and data analytics. Innovative applications include managing fleets, surveying and building roads, monitoring and mapping earthquake damage, and improving the efficiency of wireless communication networks. Trimble solutions are at work around the world and have been used everywhere from North Pole expeditions to Antarctic surveys, from re-measuring the height of Mount Everest to helping rescue miners trapped deep underground.

Smart Cities

Downtown San Jose at dusk.

Some point to connected cities as being the poster child of the IoT revolution. Many cities, including San Jose and Palo Alto, are making a big push to become smarter and more connected. A few initiatives include:

SAN JOSE

* **EMERGENCY VEHICLES:** San Jose Police and fire vehicles are now enabled with new IoT technology providing emergency responders better control over traffic lights which has improved emergency response times.
* **SMARTPHONE APP**: The city rolled out a community engagement application enabling residents to report abandoned vehicles, illegal dumping, streetlight outages, potholes, and graffiti in real time using GPS coordinates to accurately describe the location.
* **TRAFFIC DATA SERVICES**: The city created a congestion management solution which provides real time traffic and population movement information helping leaders better plan the future of transportation within the city.

PALO ALTO

* **DRONE DELIVERY:** Palo Alto is currently running a number of pilot programs allowing local companies to test the capabilities of unmanned robots delivering everything from take-out to blood samples.
* **RAIL CROSSING VIDEO MONITORING**: The city deployed a video monitoring system which uses technology to alert officials to any unusual or problematic behavior at railroad crossings across the city.
* **SMART TRAFFIC LIGHTS:** The city installed smart traffic lights which adjust real-time with traffic flow for improved safety and efficiency.

Sidewalk Labs

Sidewalk Labs has been involved in various urban tech initiatives. One such initiative was partnering with LINK NYC to convert defunct pay phone booths into free Wi-Fi hubs. LINK NYC via Wikimedia

Sidewalk Labs is a subsidiary of Alphabet (Google's parent company). The organization aims to apply their "moonshot" thinking in order to build smarter cities. Sidewalk Labs often collaborates with other organizations while trying to tackle issues such as cost of living, efficient transportation and energy usage. A few examples:

* A collaboration with the U.S. Department of Transportation to build Flow—a program that would combine real-time congestion and foot traffic information with other data to help cities create better transit routes.
* A collaboration with LINK NYC to turn New York City's unused phone booths into ad-supported Wi-Fi hotspots.
* A partnership with the government of Toronto to design a new district from the ground up that will utilize new construction methods to make housing more affordable. The city grid will be well thought out to provide a large range of transportation options (as opposed to most cities which have been built around the personal automobile.) The district will encourage innovation around energy, waste, and other environmental challenges in an effort to better protect the planet. The aim is to create a role model for future urban development worldwide.

VIEW

View Dynamic Glass is smart and will automatically adjust the window tint based on its environment. View, Inc.

Smart windows from Milpitas-based View, tint automatically in response to the sun. Every window is IP addressable and can be controlled individually or as a system, reducing heating and cooling costs as well as offering uninterrupted views (by eliminating the need for blinds). Furthermore, studies have demonstrated that occupants working in offices with View Dynamic Glass report reduced headaches, eyestrain and drowsiness which often results in higher productivity. View Dynamic Glass is installed in over 35 million square feet of buildings and growing. Installations include airports like SFO, BOS, DFW; corporate campuses like Netflix and WeWork; healthcare facilities including Kaiser Permanente and Methodist Hospitals; as well as Universities such as Duke, Colorado State, and UMass. View is now expanding on their connected platform to offer other IoT smart building applications such as security, wellness, collaboration and connectivity. To date, the company has raised nearly $2 billion in funding.

Proteus Digital Health

Artist's rendering of a microchip embedded into a pill. Dreamstime.com

Proteus, based in Redwood City, has introduced an ingestible microchip the size of a grain of sand cleared by the FDA in 2012. The tiny sensor can be embedded into prescription medication to monitor medication adherence. How does it work? Once a patient ingests the prescribed medication, the sensor becomes activated. The activated sensor then communicates with a small wearable patch on the torso sending a timestamp of when the medication was taken. The patch then communicates with the patient's mobile device which provides a record of the medication schedule to healthcare providers in real-time. According to the American Heart Association, medication non-adherence in patients with cardiovascular disease results in approximately 125,000 preventable deaths a year. Medication adherence is probably just the beginning for these ingestible microchips. Scientists are exploring additional applications using of nanosensors including sensors which would remain in the bloodstream and send an alert when they see signs of an infection or other health issue, serving as an early warning system.

Robotics

The first fully autonomous robot was created at the Stanford Research Institute (SRI) in the late 1960s with funding from DARPA (the U.S. Department of Defense Advanced Research Projects Agency). The team of 10+ scientists was led by Charles Rosen and included many pioneers in robotics, computer vision, natural language processing, and artificial intelligence. The robot featured a radio antenna, a television camera, sonar equipment, collision detection sensors, and on-board processors. The robot was tall and had a tendency to shake. Thus, he was named Shakey. While Shakey's hardware was somewhat primitive, the software architecture and algorithms were revolutionary—it combined the then-nascent industries of robotics and artificial intelligence to create a moving vehicle that could sense the world around it and use that information to determine a course of action. For example, an operator would type the command "*push the block off the platform*" at a computer console. Shakey would look around, identify a platform with a block on it, and then locate the ramp it needed to reach the platform. Shakey would then push the ramp over to the platform, roll up the ramp onto the platform, and push the block off the platform. Shakey is the great-grandfather to the technology used today in self-driving cars, military drones, voice recognition like Siri and map applications like Google Maps. Shakey is on display at the Computer History Museum in Mountain View.

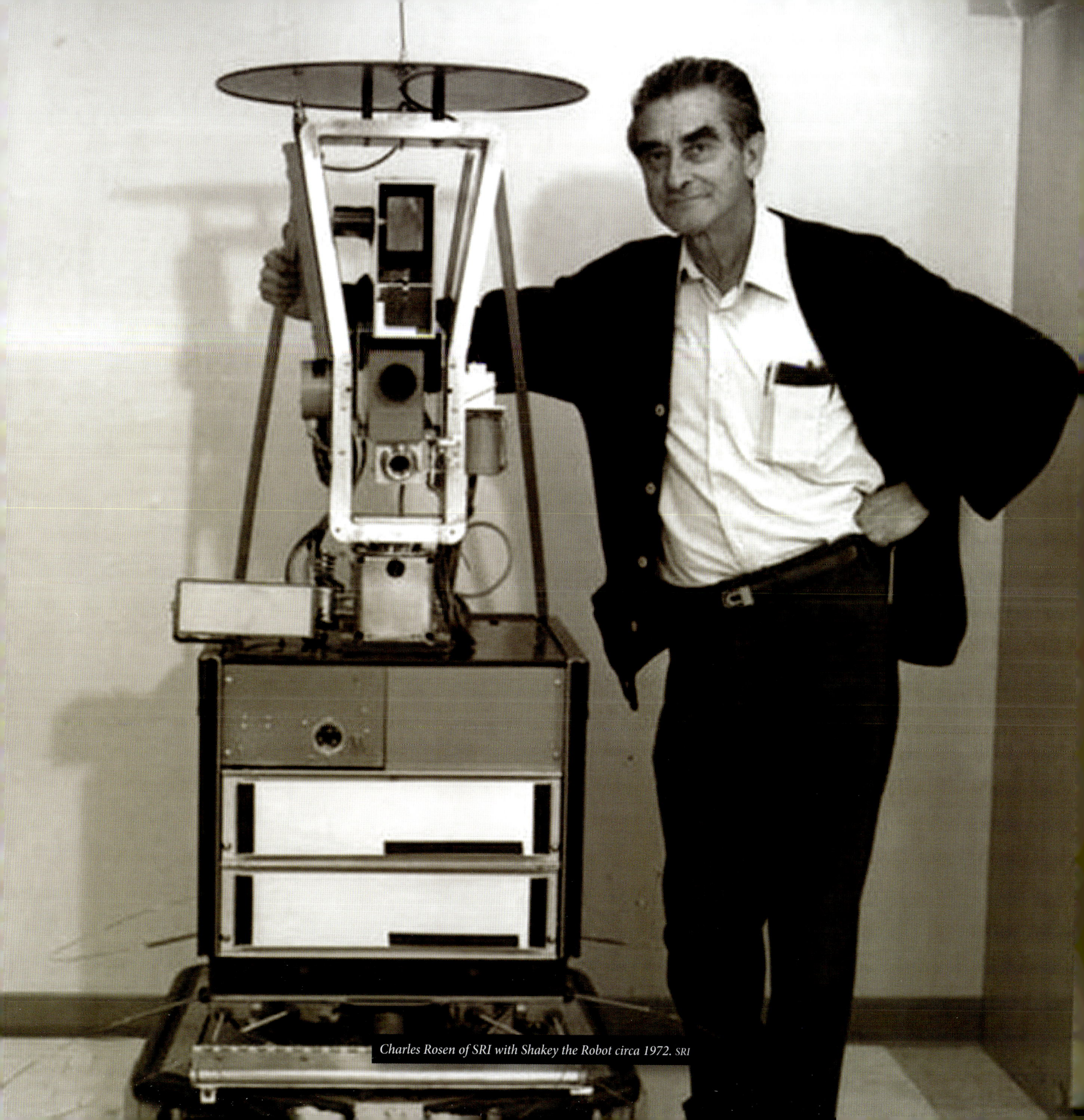

Charles Rosen of SRI with Shakey the Robot circa 1972. SRI

Intuitive Surgical

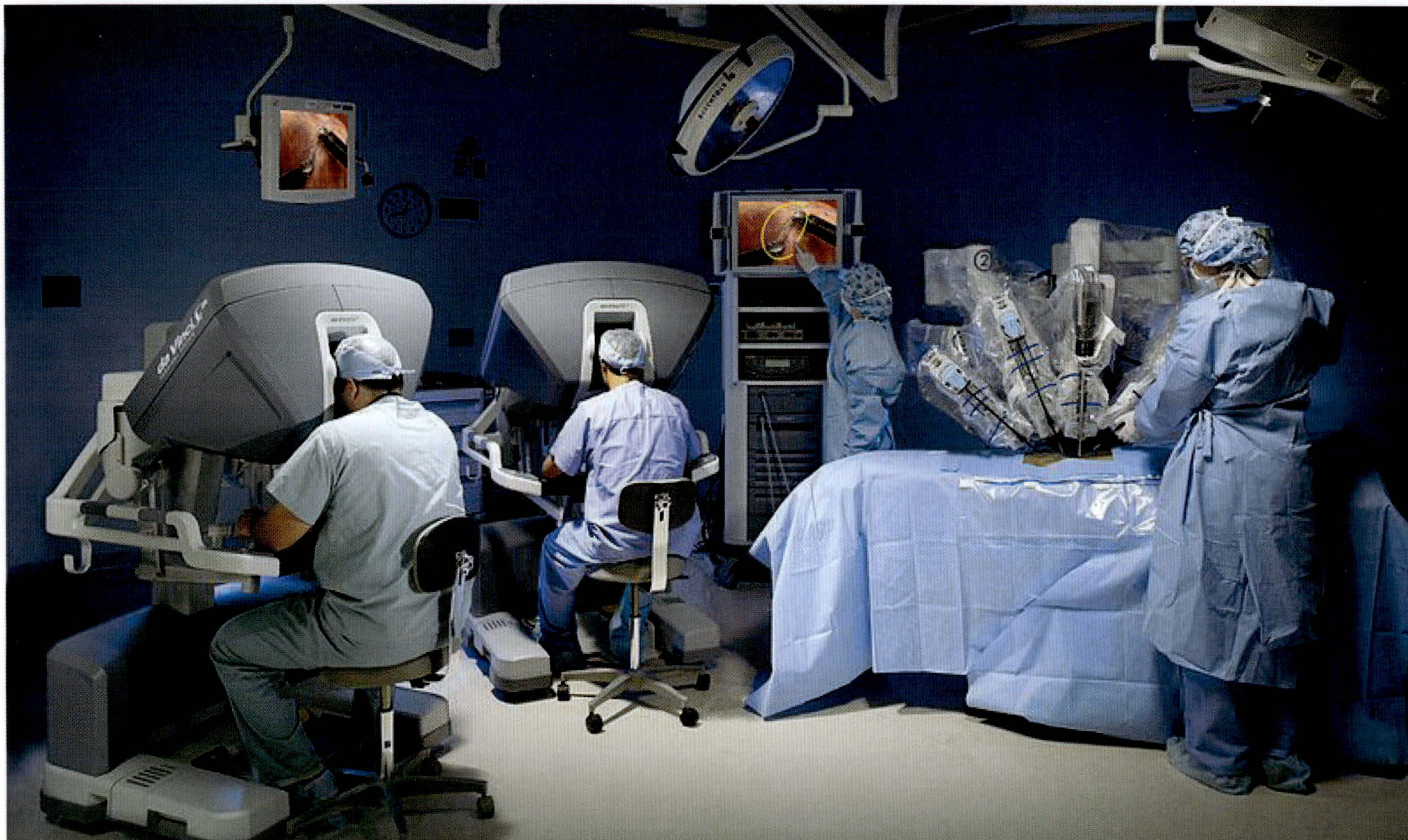

The first surgeon console dating back to 1996. Stanford Research Institute.

In the late 1980s, DARPA (the U.S. Department of Defense Advanced Research Projects Agency) asked the Stanford Research Institute to come up with a way for surgeons to operate remotely on soldiers wounded on the battlefield. The resulting product was the first U.S. FDA-approved telerobotic surgical system. In 1994, Dr. Frederic Moll and two partners purchased the intellectual property for the robotic surgical system. Moll's team raised venture capital to commercialize the robotic system and founded Intuitive Surgical Devices, now known simply as Intuitive. Today, the company's signature product is a set of robotic arms controlled by a surgeon through a separate console, although the console is typically in the same room as the patient. There are four robotic arms. Three contain surgical tools like a scalpel or a pair of scissors. The fourth arm houses a camera which gives the surgeon a 3-D view of the procedure. The surgeon uses both hand and foot controllers to operate the camera and the surgical tools. The system translates the surgeon's hand movements into more precise micro-movements of the instruments. The company went public in 2000. As of 2017, there were over 4000 units installed around the world.

Fetch Robotics

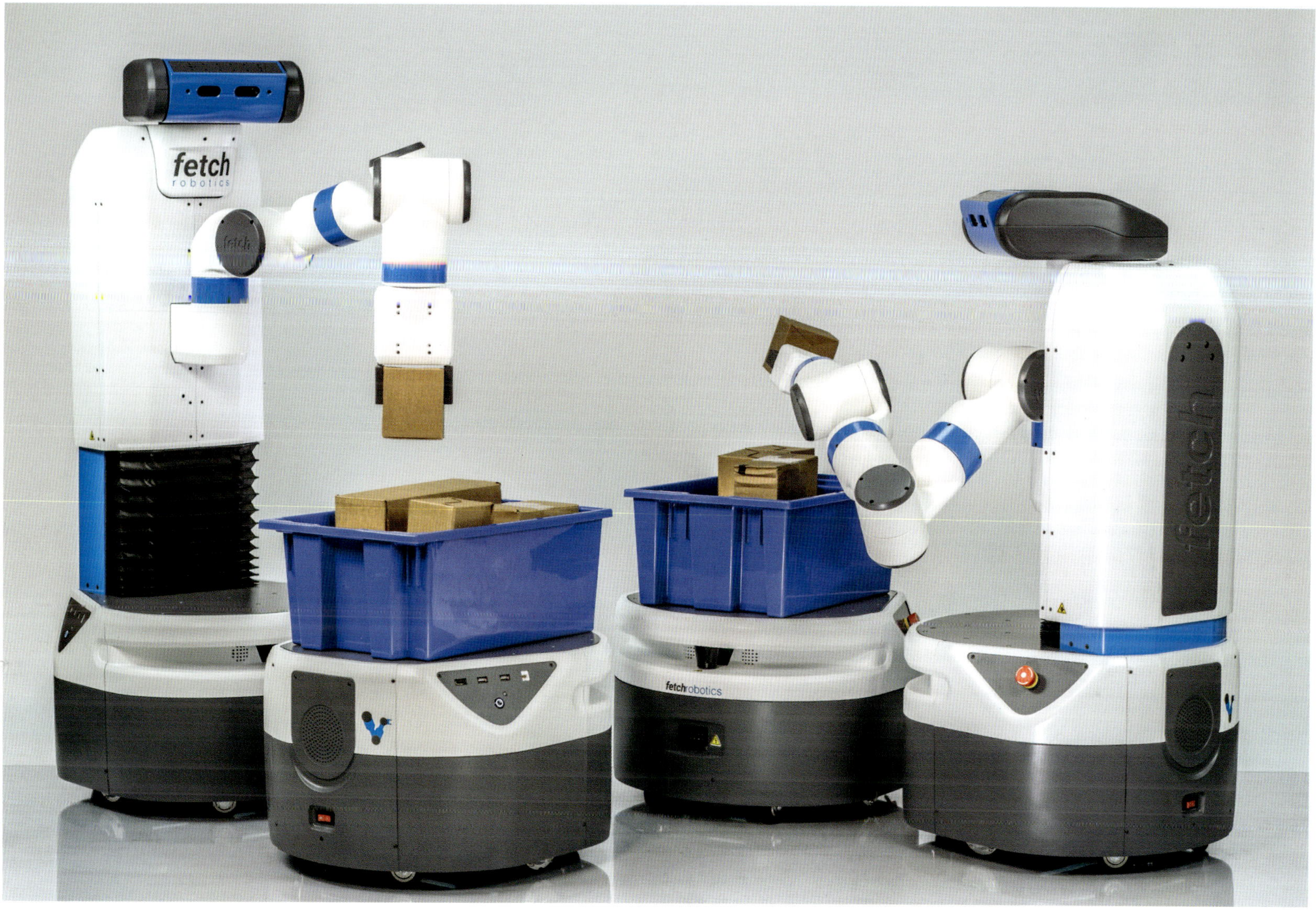

The Fetch robotic solution for the logistical market. Fetch Robotics.

San Jose based Fetch Robotics was founded in 2014 with the mission to develop robotic solutions for the warehousing and logistical market. According to CEO Melonee Wise, the exponential growth of e-commerce is driving the need for more distribution centers. However, the logistics industry is facing a shortage of people to staff them—to the tune of 60,000 unfilled positions. The Fetch robotic solution is a fleet of round robots (similar to an oversized Roomba vacuum) equipped with a cloud-based RFID (radio-frequency identification) system which allows the robots to find, track, and move anything in warehouses, factories, and distribution centers. The robots can monitor their environment making sure they avoid any unforeseen obstacles or pitfalls.

Autonomous Vehicles

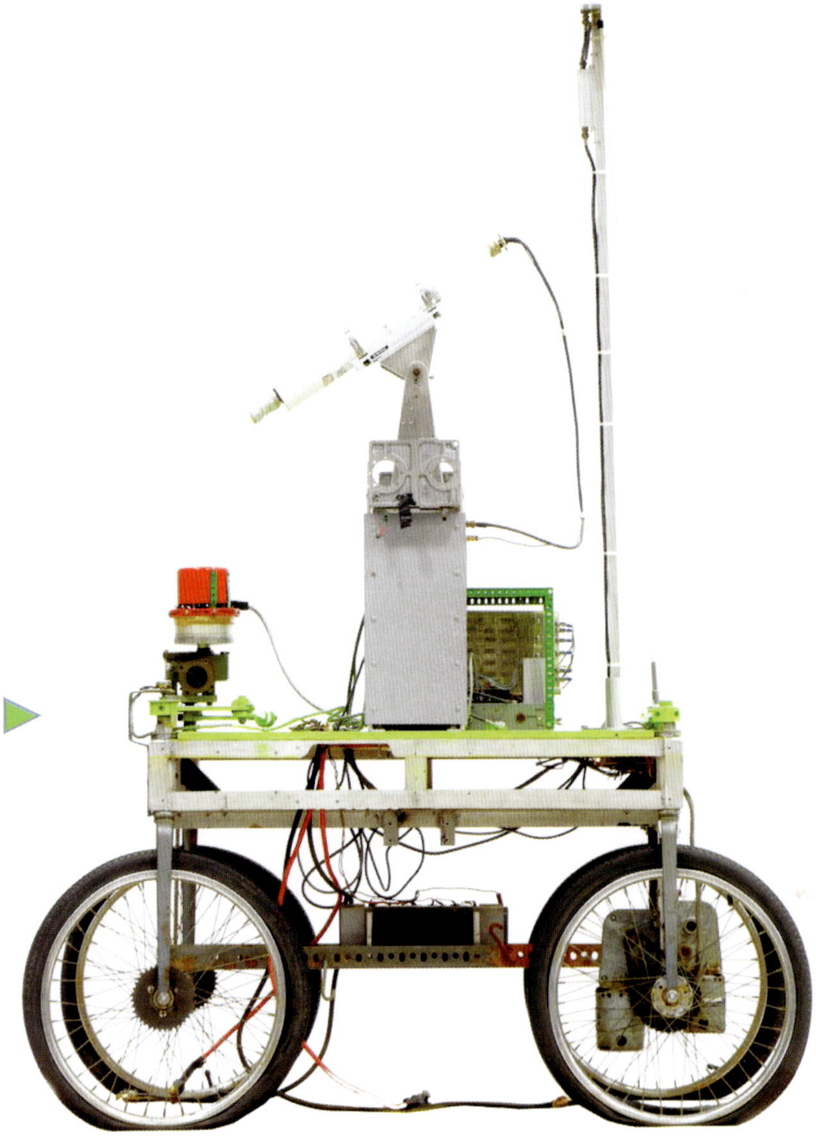

The Stanford Cart was a long-term research project undertaken at Stanford University in their Artificial Intelligence Lab between 1960 and 1980. In 1979, it successfully crossed a room on its own while navigating around a chair placed as an obstacle. *Mark Richards. Courtesy of the Computer History Museum*

The concept of a self-driving car dates back almost a century. General Motors showcased their vision at the World Fair in 1939. The idea was to create electromagnetic fields generated by metal spikes installed in the roadways. These electromagnetic fields would then act as a navigation system for autonomous vehicles. In 1977, the Japanese unveiled a self-driving car which used a camera system to relay data to a computer. The Germans improved on this a decade later by tripling the speed at which a vehicle outfitted with cameras could drive itself. Around this same time, Stanford University was working on the Stanford Cart as depicted above. In 2004, DARPA (the U.S. Department of Defense Advanced Research Projects Agency) started hosting autonomous vehicle challenges. 2005 was the first year for any vehicle to finish the course. The winning car that year belonged to the Stanford University team led by Sebastian Thrun. For many observers, these DARPA races signaled the start of the modern self-driving car era.

SEBASTIAN THRUN

Sebastian Thrun with Stanley, the winner of the 2005 DARPA Grand Challenge of autonomous vehicles. Stanford University Archives.

The father of the self-driving car, Sebastian Thrun, was born in Germany. He immigrated to the US in 1995 to join the Computer Science department at Carnegie Mellon University. In 2003 he moved out west and joined the faculty at Stanford University. He became the director of SAIL (Stanford Artificial Intelligence Lab) in 2004. Shortly thereafter, he led a team of computer scientists to build Stanley—an autonomous vehicle which won the 2005 DARPA Grand Challenge of self-driving cars. In 2007 Thrun joined Google to found and lead the highly secretive Google X labs where he worked on StreetView, Project Ground Truth and furthering autonomous vehicle technology. He left Google in 2012 to co-found online educational organization Udacity saying: "*I could be at Google and build a self-driving car, or I could teach 10,000 students how to build self-driving cars.*" Thrun then founded Kitty Hawk in 2015 with financial backing from Larry Page. Kitty Hawk aims to make personal aircraft a reality in the next five years. Today, Thrun is an adjunct professor at Stanford University, serves as CEO of Kitty Hawk and is Chairman of the Board for Udacity.

Google/Waymo

From left to right: Eric Schmidt, Larry Page and Sergey Brin in their experimental self-driving car circa 2011. Computer History Museum

The Firefly, as seen here, had no steering wheel or pedals. Computer History Museum

Google's foray into the self-driving revolution dates back as far as 2002. Larry Page, co-founder of Google, approached a few PhD students with a videotape he had captured while driving around the Bay Area and challenged them to invent a way to summarize the video with a few images. That was the beginning of CityBlock, a Stanford University research project which was sponsored by Google. In 2006 Google hired the PhD students and their work on CityBlock ultimately became Google's StreetView. Official development of autonomous vehicles began when Sebastian Thrun joined Google to spearhead the project. In 2015, Google introduced Firefly vehicles which had no steering wheel or pedals (pictured above, now retired). In 2016, Google spun out the self-driving car division into a new company under the name Waymo, short for a "new way forward for mobility." As of late 2018, the company has logged 10 million real-world miles and continues to add about 1 million miles a month to that sum. In late 2018, Waymo was the first autonomous car company to receive a permit from the California DMV to test their vehicles in the state without a human behind the wheel.

Tesla

Tesla Motors founders Marc Tarpenning and Martin Eberhard with the Tesla Roadster in 2007. Martin Eberhard

Original investor and now CEO Elon Musk with the Model S at the Fremont factory in 2011. Maurizio Pesce

Tesla Motors was founded in 2003 by Martin Eberhard and Marc Tarpenning as a response to General Motors quick entry and exit from the electric car market in the early 2000s. As documented in the film *Who Killed the Electric Car?*, General Motors produced an electric car called the EV1 in 2003 and then despite high public interest, quickly recalled (and destroyed) the entire fleet. Tesla began developing their technology by working on a premium sports car aimed at early adopters and then parlayed this into more affordable vehicles. In 2004 Elon Musk led the $7.5 million Series A round of investment and joined the board of directors as its chairman. By 2007 Tesla was suffering from an out of control burn rate and it was rumored that without further capital the company would go under. In October 2008 both Eberhard and Tarpenning left the company and Musk assumed the post of CEO. One of his first moves was to reduce headcount by 25% and shortly thereafter raised enough capital to avoid bankruptcy. In 2010 Tesla and Toyota formed a strategic partnership where Toyota agreed to purchase Tesla stock and Tesla agreed to purchase the former NUMMI assembly plant in Fremont. Shortly thereafter Tesla Motors became the first American car maker to go public since the Ford Motor Company in 1956.

Largely automated Tesla factory in the re-purposed NUMMI facility in Fremont, CA. Steve Jurvetson

The Model S was developed and built in the Fremont factory and released in 2012. It boasted a battery range of 220 miles and the ability to go 0–60 mph in 5.5 seconds. It won many awards including: 2013 Motor Trend Car of the Year, Time Magazine's Best Inventions of 2012, Car and Driver's Car of the Century, and the best cars ever tested by Consumer Reports with a score of 99/100. Tesla has been pushing their autonomous car agenda since 2013 when they introduced Autopilot, an advanced driver-assistance system. In 2016, all new cars started shipping with the sensors to enable full autonomy at a later date. As of late 2018, the Advanced Autopilot feature available in all Tesla cars offers adaptive cruise control (adjusting the speed of the car based on the other cars on the road), the ability to automatically change lanes without requiring driver steering, and to summon the car from a garage or parking spot.

Artificial Intelligence

The word Artificial Intelligence dates back to 1956 and is credited to John McCarthy, the founding father of AI. In 1965 McCarthy became the founding director of the Stanford Artificial Intelligence Laboratory (SAIL). McCarthy worked at Stanford University for 38 years until his retirement in 2000.

John McCarthy is seen here playing chess at Stanford University's IBM 7090 mainframe computer circa 1967. Stanford University Archives

AI Terminology

The rapid advance of the AI industry has created plenty of buzz words. The terms artificial intelligence, machine learning and deep learning sometimes get used interchangeably, although there are differences. Artificial intelligence, machine learning and deep learning can be thought of as a set of Russian dolls nested within each other. Artificial intelligence is an umbrella term. Machine learning is a subset of AI and deep learning is a subset of machine learning. All deep learning is machine learning, all machine learning is artificial intelligence but not the other way around. A few basic definitions:

Artificial Intelligence: An umbrella concept for all computer programs which imitate intelligent human behavior. There is a wide range of ways to simulate human intelligence and some methods are more intelligent than others. One of the most basic would be a static program of "if this then that" statements. For example, a computer tax program can be created which mimics the intelligence of an accountant with knowledge of the tax code.

Machine learning: A subset of artificial intelligence. Traditionally, computer programs were static—they did the same thing over and over until a human modified the program. In contrast, machine learning is dynamic—a program updates itself as new information comes in. Google's search algorithm is a great example. The more searches performed and the more data compiled, the better the algorithm becomes at returning a more accurate result to any given query.

Deep Learning: A subset of machine learning. A computer is programed to imitate human learning. It can start with a blank slate and learn through trial and error. Like a child, these algorithms are penalized when they make the wrong decisions and rewarded when they make the right ones. Deep is a technical term. It refers to the number of layers in a neural network (defined below). A shallow network has one so-called hidden layer; a deep network has more than one.

Artificial Neural Networks: Training computers to think like humans is achieved partly through the use of neural networks. Neural networks are a series of algorithms modeled after the human brain. Just as the brain can recognize patterns and help us categorize and classify information, neural networks do the same for computers. The brain is constantly trying to make sense of the information it is processing. To do this, it labels and assigns items to categories. When we encounter something new, we try to compare it to a known item to help us understand and make sense of it. Neural networks do the same for computers.

IBM's Watson

From Left to Right: Ken Jennings, Watson's interface and Brad Rutter competing for $1 million in prize money on Jeopardy! in 2011.

In 2011, IBM's Watson computer competed on *Jeopardy!* against the game show's two biggest champions and won. Watson was not developed in Silicon Valley, but helped propel the AI industry forward and captured the world's attention. Watson defeated both men by analyzing natural language questions and content more accurately and faster than its human counterparts. Watson wasn't allowed to connect to the Internet during the show. His training involved uploading millions of documents including dictionaries, encyclopedias, news articles, and literary works. Watson can process the equivalent of a million books per second. The computer could zero in on keywords in a clue then comb its memory for clusters of associations with those words. It would then check the top hits against all the contextual information it could glean from the clue: the category name, time, place, gender, etc. Today, Watson has been deployed into the medical field attempting to address some of the world's most pressing health challenges.

SIRI

Siri, the voice-operated personal assistant, was initially developed at the Stanford Research Institute (SRI) as part of an artificial intelligence initiative undertaken by DARPA (the U.S. Department of Defense Advanced Research Projects Agency) beginning in 2005. It was the first of its kind. In 2008 SRI spun out Siri, Inc. to commercialize the technology. At that point the talking assistant was made available for download in the Apple App Store. Apple acquired the startup in 2011 and made the virtual assistant a built-in feature in their devices starting with the Apple iPhone 4S. Apple was the first to offer voice-operated technology although their competitors quickly followed suit. Initially, most competing technologies performed far worse than Siri. Shortly thereafter, Apple split with its competitors (Google, Facebook, and Amazon) and decided not to collect data on their users. Competitors who decided to aggressively leverage their user data has made their assistants more helpful. Tim Cook has commented on this shift saying: *Advancing AI by collecting huge personal profiles is laziness, not efficiency. For artificial intelligence to be truly smart, it must respect human values, including privacy. If we get this wrong, the dangers are profound.*

DeepMind

Artificial Intelligence company DeepMind has created AlphaGo which defeated the world champion at the board game Go in 2016. The game has more possible configurations than atoms in the universe. Wikimedia.

DeepMind, founded in London in 2010 and acquired by Google in 2014, is currently considered the leader in the field of artificial intelligence research today and specializes in artificial neural networks. An artificial neural network is set up to replicate the way humans learn—essentially by trial and error. An important insight into the human brain is that different parts of the brain are responsible for processing different aspects of information. These parts are arranged hierarchically—information passes through the first "layer of neurons" and then gets passed on to the next, more senior level. For an artificial neural network to learn, it has to grasp what it has done right, what it is doing wrong. This is called feedback. The program then modifies its behavior based on the feedback received. Deepmind uses games as a way to develop and test artificial neural networks. One such game is the board game Go—it dates back 2500 years. The number of possible configurations on the board is more than the number atoms in the universe. (Wow.) Deepmind made headlines in 2016 after its program (called AlphaGo) beat the world champion, Lee Sedol, in a five-game match. The match was chronicled in the documentary *AlphaGo*.

Artificial Intelligence...a word of caution

“Our intuition about the future is linear. But the reality of information technology is exponential. That makes a profound difference. If I take 30 steps linearly, I get to 30. If I take 30 steps exponentially, I get to a billion.

–Futurist, Ray Kurzweil

“With artificial intelligence we are summoning the demon. In all those stories where there's the guy with the pentagram and the holy water, it's like—yeah, he's sure he can control the demon. Doesn't work out.

–Tesla CEO, Elon Musk

“The development of full artificial intelligence could spell the end of the human race. It could take off on its own, and re-design itself at an ever increasing rate. Humans are limited by slow biological evolution and would be superseded.

–Physicist, Stephen Hawking

“It seems probable that once the machine thinking method has started, it would not take long to outstrip our feeble powers... They would be able to converse with each other to sharpen their wits. At some stage therefore, we should have to expect the machines to take control.

–1950's AI pioneer, Alan Turing

“We could one day lose control of AI systems via the rise of superintelligences that do not act in accordance with human wishes – and such powerful systems would threaten humanity.

–Technical Fellow at Microsoft, Eric Horvitz

“I agree that the future is scary and very bad for people. If we build these devices to take care of everything for us, eventually they'll think faster than us and they'll get rid of the slow humans to run companies more efficiently.

–Apple co-founder, Steve Wozniak

SUPER-INTELLIGENCE...?

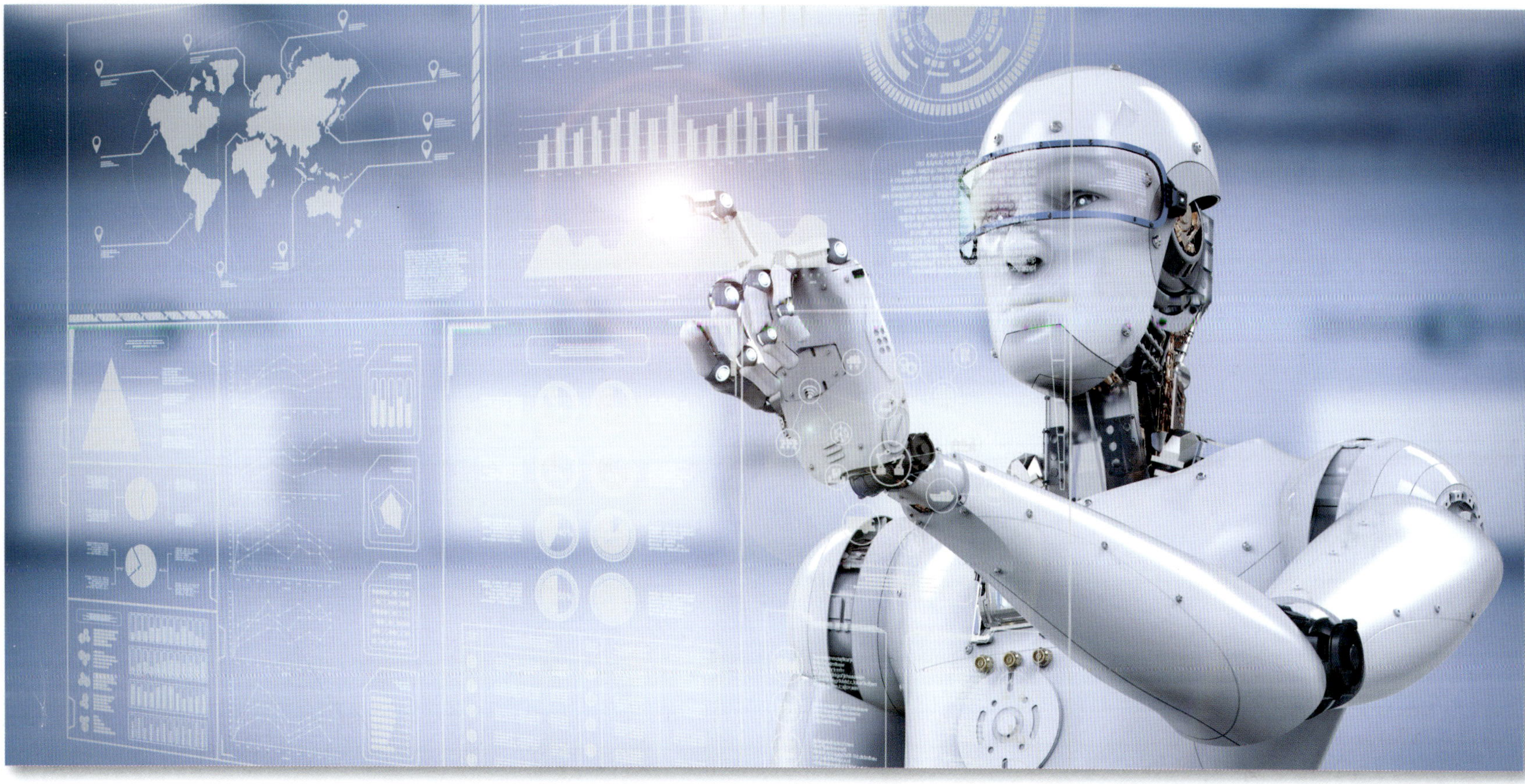

Artist depiction of what the world could look like with computers who have surpassed human intelligence. Kittipong Jirasukhanont for Dreamstime.com

Creating a machine as smart and intuitive as the human brain is the holy grail of AI research—and experts say we are getting close. But what happens once that milestone is reached? What comes next? The likelihood is that a super-intelligent system (one smarter than humans) would follow surprisingly quickly—perhaps even instantaneously. The wide reaching implications of super-intelligence are difficult to conceptualize. Ray Kurzweil calls the point at which computers surpass human intelligence as the singularity—an event horizon which is hard for us to see beyond. While it is difficult to hypothesize what the world looks like with the advent of super-intelligence, many have begun sounding the alarm bells about potential problems. Most agree that a machine that possesses technological maturity would be extremely powerful and wired for self-preservation. It would, by its very nature, actively resist any outside attempts to turn it off or otherwise prevent the completion of its programmed goal. The human race could be threatened if we somehow get in the way of those goals. Elon Musk uses the analogy of building a road. If an ant hill lays in the path of the road, it will be destroyed not out of malice, but because the ultimate goal requires it. Oxford philosopher Nick Bostrom hypothesizes that we can only survive if we ensure super-intelligence is instilled with goals that are compatible with human survival and well-being. If, when, and how that occurs is anyone's guess.

Index

Bibliography

Books:

The Silicon Valley Edge; A Habitat for Innovation and Entrepreneurship
Edited by Chong-Moon Lee, William F. Miller, Marguerite Gong Hancock and Henry S. Rowen
Stanford University Press, 2000

Understanding Silicon Valley; The Anatomy of an Entrepreneurial Region
Edited by Martin Kenney
Stanford University Press, 2000

Reconstructing Early Historical Landscapes in the Northern Santa Clara Valley
Author Alan K. Brown
Santa Clara University, Department of Anthropology and Sociology, 2005

Old Santa Clara Valley : a guide to historic buildings from Palo Alto to Gilroy
Author Phyllis Filiberti Butler
Wide World Publisher/Tetra, 1991

Technology, Entrepreneurs and Silicon Valley
Authors John McLaughlin and Carol Whiteley
Santa Clara Valley Historical Society, 2002

Silicon Valley : 110 Year Renaissance
John McLaughlin, Leigh Weimers,Ward Winslow
Santa Clara Valley Historical Society, 2008

A History of Silicon Valley
Piero Scaruffi (Author), Arun Rao (Author)
Omniware Group, 2011

Sunshine, Fruit and Flowers: Santa Clara County and Its Resources; Historical, Descriptive, Statistical
A Souvenir of the San Jose Mercury, 1895 (Hardcover)

Films:

Pirates of Silicon Valley (TV 1999)

PBS' American Experience: Silicon Valley

Something Ventured (2011)

Do You Trust this Computer? (2015)

Websites:

www.wikipedia.org

https://www.computerhistory.org

http://content.cdlib.org/

www.siliconvalleyhistorical.org/

www.npr.org/starting-up-silicon-valleys-origins

www.steveblank.com/category/secret-history-of-silicon-valley/

https://www.investopedia.com/articles/markets/051514/how-oracle-remains-relatively-unknown-software-giant.asp

https://www.seas.upenn.edu/~zives/03f/cis550/codd.pdf

https://www.titlemax.com/resources/history-of-the-autonomous-car/

https://screenrant.com/times-the-jetsons-predicted-the-future/

https://www.atariarchives.org/deli/homebrew_and_how_the_apple.php

https://www.bizjournals.com/sanjose/news/2017/04/19/facebook-terragraph-wireless-san-jose.html

https://www.inc.com/peter-cohan/an-inside-look-at-stanfords-27-trillion-turbo-cha.html

https://nypost.com/2018/07/03/facebook-owns-four-of-the-five-most-downloaded-ios-apps-ever/

https://www.forbes.com/sites/samshead/2018/04/28/google-billionaire-sergey-brin-urges-caution-on-ai-development/#240a0c9126a7

https://www.bloomberg.com/news/features/2018-06-08/tesla-model-3-photos-of-elon-musk-s-factory-in-fremont